Moho Motion

Moho Motion

How the Atlantic Widens

C. James Blom

ISBN: 1514659964
ISBN 13: 9781514659960
Library of Congress Control Number: 2015910115
CreateSpace Independent Publishing Platform,
North Charleston, South Carolina

Description of **Moho Motion**

These are the musings and sketchings of an old geologist informally proposing how the Atlantic Ocean gets wider. **Laymen**: this is pitched to you. **Geoscientists**: your initial knee-jerk reaction might be "blithering blasphemy"! Later these ideas may grow on you.

One hundred fifty+ million years ago the Atlantic was a lineup of fresh water lakes, and now it is an ocean 3000 miles wide. Down the middle of the Atlantic a submarine mountain chain overlies miles-thick volcanic rock. Continuing eruptions of volcanic rock below those mountains add an inch a year to the edges of two of the earth's plates which meet there. The Moho is a world-wide interface from 5 to 30 miles below the earth's surface. Velocities of sound are slower above and faster below the Moho. Oceanic crust and continental crust float on the Moho. The material below the Moho is usually stable, but if its confining pressure drops, that material locally transitions to lava. This is a reversible change of state, involving the **release** of heat when lava is created under the expanding Atlantic, and involving the **acquisition** of heat and energy by slabs of transitioning oceanic basalt descending hundreds of miles around the edges of the shrinking Pacific.

Pressure-depth graphs are essential tools for understanding the new theory. A byproduct of using this discipline is a more detailed interpretation of the May 18, 1980, eruption at Mt. St. Helens. There were two fluid systems in play that day: water and magma. Initially there was a 5.2 earthquake. This triggered a large landslide, which unleashed a massive underground steam explosion, lasting a minute or two. This reduced the pressure restraining the magma (lava), which then erupted.

Contributors
Cover by Tatiana Vila

Categories
Science and Mathematics
Earth Sciences
Geology

Language
English

Search keywords
Mid-Atlantic spreading, mechanics of plate tectonics, Moho shape, stress in spreading centers, subduction of oceanic crust, plume problems, 1980 Mt. St. Helens eruption

Printed by Create Space,
an Amazon Company

Available from Amazon.com,
Create Space.com and other retail outlets

Moho Motion: How the Atlantic Widens, an ebook,
was published in May 2015

Dedication

This work is dedicated to my loving friend and wife of 62 years, Olly.
You are an inspiration and the backbone of our family. Thank you, Darling.

Table of Contents

List of Figures

One

Introduction

Moho Motion. A little alliteration never hurts when contriving a catchy title. Catching the interest of geoscientists **and** the general public is the objective. Geoscientists will be interested parties because "Moho" is one of their favorite scientific terms. But the general public? People are curious about the mystery of how the earth's crust is shifting about. The Atlantic is getting wider, and the Pacific is shrinking. It's a fantastic story which has been unfolding rapidly over the past few decades. Here you will find a few new twists which are quite different from established dogma.

Welcome to the world of geologic controversy!! Here's something to tease the imagination. This is an alternate theory of how the Atlantic expands. Not many earth scientists will embrace these concepts just yet, but let's be a little daring and examine something new and promising.

Plate tectonics is about the jig saw puzzle of huge portions of the earth's surface which move about at the speed of the growth of the nail on your big toe. Each plate is separate, and each butts up against the plates next door.

Substantiating a new geologic concept is easier when dealing with a relatively simple geologic province. As the old joke goes about the drunk explaining one dark night why he is searching for his lost car keys under the street lamp: "That's where the light is". So, where is a simple geologic province where plate tectonics is involved?

The prime candidate is the submarine range of mountains running the length of the Atlantic Ocean, the mid-Atlantic ridge. This ridge lies under two miles of water, which explains why it is very lightly explored. So if exploration there is sparse, why is the geology there deemed simple?

Let's put it in perspective. More than two thirds of the earth is covered by oceans. The rocks beneath the oceans are mostly volcanic, giving up their secrets to man only around the edges where petroleum resources are sought, or where oceanographers have barely scratched the surfaces. Our moon, Mars, Venus and moons of Jupiter and Saturn have been extensively observed and measured by NASA's remarkable spacecraft. On the other hand, worldwide mapping of the shape of the earth's ocean bottom was only undertaken to satisfy the curiosity of the major navies which wanted to hide their submarines. What's deeply hidden below the bottom of the oceans is the subject of theories. Not too many facts.

The shape of the ocean bottom is commonly measured by echo sounders operated from floating vessels. The abyssal plains on either side of the mid-Atlantic ridge average more than three miles deep, more than a mile deeper than the top of the ridge. For thousands of miles along the bottom of the mid-Atlantic these abyssal plains, and the ridge separating them, exist without much variation in bathymetry. On the plains the thin dusting of recent sediment thickens very gradually away from the ridge. This thin dusting seems to show little or no disturbance, so we conclude that the ocean bottom has been stable, and therefore the rocks below have remained undisturbed. Earthquakes under the Atlantic are limited to the mid-Atlantic ridge, suggesting that under the abyssal plains the earth's crust is inactive.

Under the Atlantic is relatively simple, quiescent geology extending for millions of square miles with only one active zone: the mid-Atlantic ridge. This is a likely place to look for a mechanism causing the earth's plates to move away from the ridge. And perhaps insights gained would be transportable to other sectors of the 30,000 miles of mid-oceanic ridges, meandering around the earth.

For the layman, geologic publications are heavy slogging. In geologic literature, an author will bolster his case for incremental improvements to geologic knowledge, citing numerous references to the works of other geoscientists, sometimes using insider terminology and an alphabet soup of abbreviations. But in the case of **Moho Motion**, the intent is to rearrange the known facts to support a new theory. The new theory is only marginally based on existing dogma, so extensive references to geologic literature are irrelevant.

Here are presented new geological concepts expressed in layman's terms, without the complications of the formal scientific literary style. The ideas presented are not incremental additions to the state of the art, but rather inquiry in a new direction. Although this style is informal, the concepts presented are serious stuff! In **Moho Motion** no new measurements, no results of research and no improvements in accuracy are the subjects. Here are presented theories which have not been tested. Thus, an informal presentation, suitable for geoscientists, is also appropriate for the general public. It will be up to geoscientists to devise tests and make new measurements to probe and question the ideas expressed here.

Two

Logic for New Theory

Geology has its exact and inexact applications. Minerals are exactly known, from chemical content to crystal structure. But inexactness dominates geology. From studying a few patches of rock sticking up above the ground, a geologist uses his experience and training to concoct a most likely picture of what's hidden below. In the news you could follow the measurements by the Mars Rover as it gathered information for geologists to fathom the nature and history of that distant planet. **Moho Motion** is another inexact application. The goal is to make our understanding of the mid-Atlantic ridge less inexact. And insights may be transportable to other areas of the world.

To concoct an improved theory of the mechanism for plate tectonics here on earth, a logical case should be constructed. Start with **facts**, identify **assumptions,** then make **judgment calls** and arrive at **conclusions.** The objective is to arrive at a most likely explanation of how two of the earth's plates move east and west away from the mid-Atlantic ridge.

The mechanism of plate tectonics occurs miles deep below the ocean bottom, where no man will ever go. So these ideas are certain to be controversial. But a carefully assembled story, recognizing the shortcomings in current dogma, should challenge geoscientists to think again.

But before we plunge into this process of derivation, let's set the stage. To put new concepts in perspective, we review the history of the older idea of continental drift and the newer theory of plate tectonics. Following that will be a critical review of current popular concepts of plate tectonics, warts and all. Next will be a review of the actual known facts pertaining to the plate tectonics of the mid-Atlantic. There aren't so many actual facts.

Next we begin the process of contriving a new theory, step by step. Starting with a basic premise, questions arise and answers are hypothesized. These answers lead to more questions. More answers. The objective is to balance the resulting theory against the facts. Soon we arrive at a hypothetical explanation for the mechanism responsible for generating the oceanic crust which now covers two thirds of the earth's surface.

Some related topics require explanation. A thorough explanation of the relationships of pressure and depth leads to understanding of the stress in rocks sliding off the mid-Atlantic ridge. The pattern of stress defines the likely symmetry of the deformation of the rocks there. Next there is energy entrapment along the spreading center. The causes for puzzling imperfections in the magnetic stripes in the basalt under the Atlantic are considered.

Around the shrinking Pacific, oceanic crust is deflected down under the surrounding continents. It is proposed that the cause of powerful earthquakes in this sinking crust is a direct result of the earth's daily rotation.

Large volcanic eruptions occur in the mid-Atlantic in two primary formats: cones and flows. Lava chooses the format requiring the least work to escape from captivity, and to buoyantly rise.

Three

Conclusions

Your curiosity may be piqued. Rather than first having to wade through all of the aforementioned studies, it may be useful for you to know the conclusions before encountering the rest of the text. Thus you will have an abstract of the ideas presented so you can judge, as you go along, whether the arguments presented are sound and the conclusions justified. This preview will only present the bare conclusions, not the facts, judgment calls or assumptions. Many of these conclusions are old hat to geoscientists. Laymen, on the other hand, may need some background information to understand the arguments abstracted here. So these conclusions have been "laymanized".

The conclusions are divided into thirteen groups.

- **THE MOHO**
  - The Moho is an abrupt change in the velocity of sound at the base of the earth's crust. Slower above and faster below, as detected by studying the noise from earthquakes. The Moho occurs worldwide at a depth of a few miles under the seas and deeper under the continents.
  - The Moho marks the boundary between the earth's crust and the material below the Moho.

- **BASALT ROCK UNDER THE OCEANS**
  - The Atlantic Ocean is spreading while the Pacific shrinks.
  - More than two thirds of the earth's surface is covered by ocean. With minor exceptions the rock type under all oceans is basalt resulting from cooling of lava generated at mid-oceanic spreading centers. Around the edges of the Atlantic thick sediments, eroded off the land, cover the oldest basalt, generated when spreading started.
  - The age of basalt under the oceans today is no more than a couple hundred million years. Chicken feed compared to the 4.6 billion years age of the earth.
  - Unique sea life is observed from diving vehicles exploring the mid-Atlantic ridge. At the base of this food chain are bacteria feeding on the chemicals in hot fluids cycling up from volcanic activity below the seabed. These volcanic conditions have existed with sufficient continuity to sustain evolution of unique sea life along the mid-Atlantic ridge.

- The earthquakes generated in the spreading centers under the crest of the mid-Atlantic ridge show that mid-oceanic basalt at those locations is occasionally stressed beyond the breaking point.

- ## NATURE OF THE MATERIAL BELOW THE MOHO

  - The three layers comprising the solid earth are: 1) crust, 2) material below the Moho and 3) core.
  - The material below the Moho is a mat of compressed crystals of various minerals in a dense, high-pressure and high-energy "soup" of ions, connected in the mat by molecular forces.
  - The pressure in the "soup" is maximum, i.e. equal to the sum of all weight (force) from above to that depth. In geological lingo, it's at lithostatic pressure.
  - The material below the Moho is a little like silly putty: its nature depends on whether it is measured in milliseconds or years. In split-seconds, waves of sound traverse the material as if it were a solid rock. Over time it acts like a mushy viscous fluid.
  - The earth's plates under the Atlantic, slide on the frictionless, dense, material below the Moho at the speed of fingernail growth.
  - The earth's crust, oceanic and continental, floats on the material below the Moho.
  - This material below the Moho has no strength. It does not break, and hence there are no earthquakes generated there.
  - Around the shrinking Pacific there are curtains of deeper earthquakes (Benioff zones). These quakes do not occur in the material below the Moho, but rather in oceanic crust which was deflected down after bumping into the plate next door. The oceanic crust is sliding down into the stationary mushy material below the Moho.
  - A recent publication indicates that below northeastern Japan, where the Pacific plate is deflected down under Japan, the Benioff zone of hypocenters comprises two six-miles thick layers separated be a 12 miles-thick dead zone. According to the theory presented here in **Moho Motion,** the upper layer of hypocenters would be the subducting oceanic crust. The lower layer of hypocenters could be phantom: the work of some acoustic chicanery. If the lower layer of Benioff hypocenters proves to be real, that spells trouble for the theory presented here.
  - Isostasy, or the change of elevation of the earth's surface in response to glacial or other loading, is the result of crust floating buoyantly on the material below the Moho.
  - New oceanic crust occurring mid-Atlantic is formed by lava cooling and becoming solid after leaving spreading centers.
  - New crust attaches to, and becomes part of each of the two plates departing east and west from the mid-Atlantic ridge.
  - The two mid-Atlantic plates are elastic solids with greater strength and rigidity in the cooled upper thousands of feet.

- **GRAVITY SLIDING OF PLATES**
  - In the Atlantic the two plates slide downhill east and west off the humped-up Moho under the mid-Atlantic ridge.
  - The Moho is higher under mid-Atlantic spreading centers than below the mile-deeper abyssal plains to the east and west. Gravity sliding down this elevation difference is the source of energy which drives plate movement away from Mid-Atlantic spreading centers.

- **HEAD OF THE MATERIAL BELOW THE MOHO**
  - Borrowing a term from hydraulics, the **head** of the material below the Moho is its potential energy, the height that viscous material would rise in a virtual standpipe, until it could rise no higher.
  - Under the mid-Atlantic ridge the Moho has remained higher than under the abyssal plains since the ridge formed, about 150 million years ago. This means that below the Moho there is higher head under the ridge than under the plains.
  - If the head in the material below the Moho is higher under the ridge, it follows that elsewhere its head is lower, so there is a dynamic system in operation below the Moho.
  - It is not necessary to know how the mechanism works. Just recognize that it does. The mechanism has fueled spreading of the Atlantic by keeping the Moho higher under the ridge ever since the birth of the mid-Atlantic ridge.
  - The mechanism controlling the head in the material below the Moho need only cause the material below the Moho to rise under the broad mid-Atlantic ridge a very small fraction of an inch annually to provide enough material 1) for conversion to the new basalt generated annually in the spreading center, and 2) to keep the Moho a mile higher there.

- **FORMATION OF LAVA**
  - The material below the Moho is stable at lithostatic pressure.
  - The material below the Moho will spontaneously liquefy to lava for as long as its confining pressure is reduced locally below its stability condition (lithostatic loading).
  - Mid-Atlantic basalt has the same chemical composition as the denser material below the Moho. These are two different states with identical chemical composition, and the conversion from one state to another is reversible.
  - Along the spreading centers under the Atlantic ridge, the packet of new solid basalt is pulling apart west and east as the two plates slide off the mounded Moho. Lava is formed just below the Moho wherever the total vertical stress on the Moho is less than the stable condition for the material there just below the Moho.
  - Two conditions cause temporary reduction of pressure on the Moho. First is the active zone of "pull-apart". Fault-bounded horizontal prisms of basalt oriented north-south, support one another, temporarily shielding some places just below the Moho from the full vertical load. Second, the upper thousands of feet of oceanic crust are cooled and rigid, so when the newly-formed

plates slide apart, the two sides may lean towards each other, thereby forming an arch with re-distributed vertical stress within the rock below.

- Earthquakes in the mid-Atlantic spreading centers confirm that basalt fractures while spreading is occurring. Presumably these faults define horizontal, elongate blocks, oriented north-south, which settle by gravity in the spreading zone.
- Liquid lava, being less dense, works its way up into the solid basalt until the back-pressure builds sufficiently to return the pressure immediately below the Moho to its stable condition. Then the flow of lava ceases. Thus the "lava breeder" dies.
- The location of each "lava breeder" may be offset miles east or west from the line on the ocean floor separating the two plates.
- Further away from Atlantic spreading centers, where "lava breeders" don't exist, basalt has become "healed" and no longer participates in the "pull-apart".
- Major basalt eruptions worldwide occur in three environments: mid-oceanic ridges, flood basalts (onshore and offshore), and the Pacific Ring of Fire. A case is made that all three modes of major volcanism result from the same cause: locally reduced vertical pressure on the Moho.

• RELEASING HEAT OF LIQUEFACTION
- Under the Atlantic, the **heat of liquefaction** is released, when lava is generated. Under the Pacific the **heat of liquefaction** is introduced into subducting basalt.
- Slabs of oceanic crust around the Pacific rim spend millions of years during hundreds of miles of descent at the speed of "finger nail growth ", while experiencing earthquakes in the Benioff zones. Below about 450 miles subsea, there are no earthquakes. This sequence is interpreted to mean that around the edges of the Pacific, the subducting basalt slabs slowly transforms to the other state, the denser material below the Moho. The transformation is complete at 450 miles depth. Outer layer by outer layer, the **heat of liquefaction** was acquired by subducting oceanic crust, changing it to be the same as the surrounding material below the Moho.
- "Lava breeders" in the material below the Moho release so much heat that lava exits the "lava breeder" at 2100 degrees Fahrenheit.
- Dissipation of the released **heat of liquefaction** under the mid-Atlantic ridge is complex, involving thermal conductivity, easiest path for the lava to rise, complications from the presence of mobile sea water, storage of heat as super-heated water and movement of that hot water.
- As solid basalt cools through the Currie point (about 1060 degrees Fahrenheit) the magnetic minerals in the basalt take on the same magnetic orientation as the earth's magnetic field. The polarity of the earth's magnetic field has flipped from north to south many times since the Atlantic commenced spreading.
- As the Atlantic broadens, successive periods of "north" or "south" magnetic orientation would be expected to have a nice piano-key appearance on a map of magnetic intensity. Actually some

of the stripes seem indistinct and are often internally inconsistent. These complications may arise from: a) off-center locations of lava breeders a few miles east and west from the line of separation of the two plates, b) mobile hot water distributes cooling and reheating, complicating the whereabouts of the Curie point, and c) inaccuracies of measuring the magnetic field through two or three miles of sea water.

- **CYCLE OF OCEANIC BASALT**
  - Consider the cycle of mid-oceanic basalt. 1) Starting under the Atlantic as the stable material below the Moho, 2) there is a local reduction of the confining pressure immediately below the Moho which destabilizes that material, 3) which then liquefies and heat is given off, 4) oceanic crust forms as lava solidifies, and the Atlantic widens. In the Atlantic there has been no loss of oceanic basalt; all of it remains incorporated into the two departing plates. For the other half of the story we switch to the shrinking Pacific, where oceanic crust is being consumed. 5) a water-logged slab of oceanic basalt is deflected down ("subducted") into the material below the Moho, until 6) the relaxing bottom side of the bending slab is the site of minor volcanism (the Pacific "Ring of Fire"), 7) the slab continues subducting until 8) all the **heat of liquefaction** is restored, so that 9) all the subducting oceanic crust eventually transforms into stable material below the Moho. 10) The basalt has gone full cycle (combining Atlantic and Pacific activity).

- **DEEP EARTHQUAKES**
  - Around the edges of the Pacific, where oceanic crust is deflected down under the continental crust, the slab of oceanic crust slides down hundreds of miles into the stationary mush. This movement occurs in the channel cooled by earlier passage of that crust. Some of the world's most powerful earthquakes occur in the channel at depths as great as 450 miles. So what causes these powerful shocks? At low latitudes, as the oceanic crust slides a few hundred miles deep, the radius of daily rotation around the earth's axis becomes significantly less. The slab "tries" to maintain its momentum by pushing to the east against the stationary, viscous mush, and when it cannot get relief fast enough, the brittle basalt breaks, causing these earthquakes.
  - Under northern Japan earthquakes seem to be localized along the **upper** surface of the slab of subducting oceanic crust, and the shape of the subducting crust is concave downwards. Under the eastern Pacific, where the shape of a subducting slab is concave upwards, earthquakes should generally be localized toward the **lowe**r surface of the slab.

- **ICELAND**
  - Iceland is a chunk of orphaned continental crust left over in the mid-Atlantic (a mini- plate) as the spreading centers zippered their way north. Well, not a clean zipper action because the "zipper"

in the Atlantic is frequently offset by transform faults. The tracks of the spreading centers split and encircled Iceland.

- On the east and west sides of Iceland, lava is released just below the level of the **oceanic** Moho. Half of the mid-Atlantic spreading occurs adjacent to the west side of Iceland, and half adjacent to the east side. Some of the lava leaks up through the upper part of the block of continental crust onto the surface of Iceland.
- Similar to Iceland, other smaller mini-plates of older continental crust occur south of Iceland along the crest of the mid-Atlantic ridge.

• FLOOD BASALTS AND HAWAII

- Onshore there are areas of thick volcanic rock in various places around the world (Columbia basalts in eastern Washington, Deccan Traps in western India, etc.) which have chemical compositions very similar to the mid-oceanic basalt from the Atlantic and the Pacific.
- Flood basalts may occur when a moving plate rides up onto an upwelling of the Moho. This tends to split the moving plate apart, on top of the upwelling. The split occurs in the direction of plate movement. The effect of Poisson's ratio will keep the two sides of the split in contact. A lava breeder, oriented parallel to the direction of plate movement, will occur locally where vertical pressure on the Moho decreases to less than the stability value. This may happen under oceanic crust as well as under continental crust.
- Mauna Kea on Hawaii is the tallest volcano on earth, extending up three miles above the abyssal plain, and almost another three miles above sea level. In all, six miles tall! Lava rises buoyantly because it is less dense than the surrounding solid basalt. The chemical content of this basalt is very similar to all basalt generated at mid-oceanic ridges worldwide. At the source of Hawaiian lava, maximum pressure in the lava must be equal to the surrounding pressure in the material below the Moho. The depth of the source must be several miles lower down below the depressed Moho in order for the lava to have had the energy needed to buoyantly rise to the top of Mauna Kea.
- Under Hawaii could there be a very slow-speed cyclonic disturbance in the material below the Moho? In the vertical vortex, there would be a reduction of pressure which could be sufficient to destabilize the material below the Moho, and cause intermittent formation of lava there.
- The submarine track of Hawaii's predecessor volcanos, extends far to the northwest, suggesting that the Hawaiian "hot spot" (vortex?) has been active for more than a hundred million years under the moving oceanic crust. There is a kink in the direction of the submarine track, indicating an abrupt change in the direction of crustal movement over the stationary source of lava.

- **PURSUING A GEOLOGICAL HOLY GRAIL
  OF CORING ROCK BELOW THE MOHO**
  - The Russians spent 17 years drilling the Kola well, which is (as of 2015) the world's deepest, in northern Russia near Norway. They drilled to 40,230 feet, but unexpectedly higher temperature caused abandonment in 1997.
  - The American Mohole Project (1958-1966) intended to drill to the Moho offshore where the earth's crust is thinner, but government funding was lost.
  - In 2012 at the Offshore Technology Conference, technical plans were proposed for coring the Moho from a drilling vessel located offshore in water depth exceeding 10,000 feet.
  - If the theory proposed in **Moho Motion** is correct, representative cored material from below the Moho would turn into molten lava while being withdrawn from the borehole. Disappointing, to say the least, to earth scientists expecting a core of solid rock.

- **AND IN ADDITION TO ALL THE ABOVE THERE ARE
  INTERESTING SIDELIGHTS IN MOHO MOTION**
  - Using pressure-depth graphs to describe buoyant fluids (petroleum, steam and magma) trapped in rocks, and to estimate stress distribution in buoyed rock
  - The underground steam explosion initiating the 1980 Mt. Saint Helens eruption.
  - Estimating the head ("height in a standpipe") of the material below the Moho.
  - Graphic solutions for "floating on the Moho".
  - A geologist's prejudice.
  - Several reality checks are proposed for the theories in **Moho Motion.**

❖   ❖   ❖

Four

Continental drift

Alfred Wegener in 1915 was the first to seriously propose that the matching coastlines of the Atlantic had split down the middle and drifted 3000 miles apart. Naturally, this imaginative concept of continental drift attracted much apathy, derision and little support from the geologists of the day. The idea was bold, but there were no data on which to hang your hat.

In 1872 a route was sought for the trans-Atlantic cable, and scientists aboard the HMS Challenger found that the sea bottom was shallower in the mid-Atlantic.

❖   ❖   ❖

Five

Plate Tectonics

By the 1960s Admiral Harry Hess' and others' oceanographic studies showed that the bottom of the Atlantic sloped away from a long, active volcanic mountain chain in the middle of the Atlantic. In 1963 geophysicists Fredrick Vine and Lawrence Morley realized that million-year reversals in the earth's magnetic field were "frozen" in the magnetic orientation of minerals in the basalt under the oceans. These magnetic stripes on either side of the mid-oceanic ridge are strikingly symmetrical. When the magnetic pole was in the north, the newly formed basalt stripe down the middle of the mid-Atlantic ridge had a northerly magnetic orientation. After the flip-flop of the magnetic field, new crust being added down the middle of the ridge had the opposite orientation. The symmetry of the stripes, from one side of the spreading centers to the other, implies that the western plate moves west at about the same rate as the eastern plate moves east.

These new insights were persuasive, and plate tectonics soon became accepted as the new paradigm by the geoscientist community.

Under the oceans the earth's crust is a layer of dense basalt, several miles in thickness (oceanic crust). Continental crust under the world's seven continents is much thicker and less dense.

A. SPREADING CENTERS

More than 30,000 miles of mid-ocean ridges have been mapped under the oceans worldwide. To appreciate this astounding length, note that the distance around the earth is about 25,000 miles. These centers of submarine volcanic activity form the meandering crests of submarine mountain ranges. Beneath these crests new oceanic crust is being created. At each spreading center the two plates slide apart as much as inches per year. Released lava intrudes up into solid basalt, taking advantage of directions of weakness in the basalt. The lava congeals into solid basalt. The new oceanic crust being formed attaches to each of the two plates moving away from a spreading center. At the far edges of each plate there is a collision with another plate. The North American plate, from the middle of the Atlantic to the coast of California, is almost 5000 miles wide, or about a fifth of the distance around the world. Offshore from California is the Pacific plate.

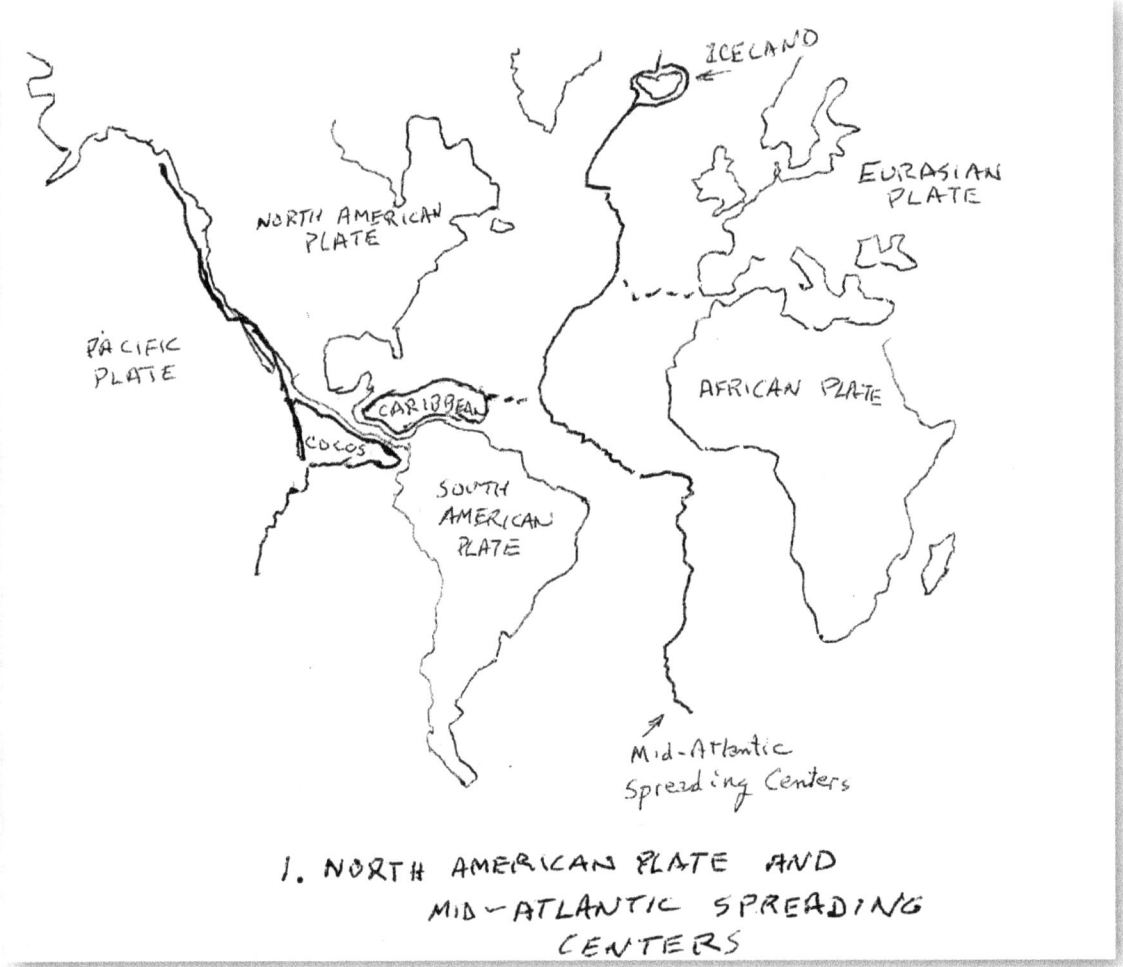

Figure 1. Location of mid-Atlantic spreading centers. Individual spreading centers are offset from each other by transform faults.

The submarine bathymetry of the Atlantic ridge is about a mile higher than the older oceanic abyssal plains on either side of the underwater mountain range. The crest of the range is covered by an average depth of two miles of water. Water depths over the abyssal plains average more than three miles

Google Earth, in its online display of the earth, includes ocean bathymetry. It is wonderful! You can sit at your computer and "fly" in the oceans worldwide. Coordinates and water depths are displayed for all locations. You will observe the trend of the mid-Atlantic ridge, some of its frequent offsets, and uneven topography with more than a thousand feet of relief. However Google's cartoon of the ocean bottom does not agree well with the displayed depths. The artist may have been unduly influenced by the magnetic stripes and the paradigm of plate tectonics.

Oceanographers use submarines to explore the mid-Atlantic ridge. "Black smokers", spew mineral-rich hot black water from natural pipes rising many feet above the seafloor. These pipes precipitate from chemicals in the "black" water. Unique sea creatures live in the poisonous habitat of submarine volcanism there. These creatures are accustomed to hot water laden with chemicals, which cycles up from cooling basalt. Unique single-celled organisms are at the bottom of the food chain here. Moving up the local food chain there are more advanced forms of life, but these creatures live exclusively near submarine volcanism. Humans would call this a very toxic environment, but it suits the unusual critters just fine, and they cannot survive long without it.

B. EARTH'S PLATES

Plates may or may not include land. The outdated term "continental drift" gives the impression that only continents (land) are involved. Incorrect! Some mini- plates are entirely mid-oceanic basalt. Plates grow as volcanic eruptions add basalt to the edge of each plate at spreading centers. Around the shores of the continents, sediments eroded from land spill out onto the oceanic basalt.

C. CRUST BEING DESTROYED

In the expanding Atlantic, new oceanic crust is being **formed.** In order to follow the **destruction** of oceanic crust we have to shift our attention to the shrinking Pacific. Continental crust is thicker and less dense than the relatively thin oceanic crust generated at submarine spreading centers. So when a plate of thin oceanic basalt collides with a thick plate of continental crust bordering the Pacific, the slab of oceanic basalt is deflected down. Oceanic crust flexes a second time when it straightens out in passive, stationary material below the Moho, after it is no longer in contact with the continental crust. The descending oceanic crust is traced by its "snap, crackle, pop" seismic signature which dies out at 450 miles depth, signifying the last of its death throws.

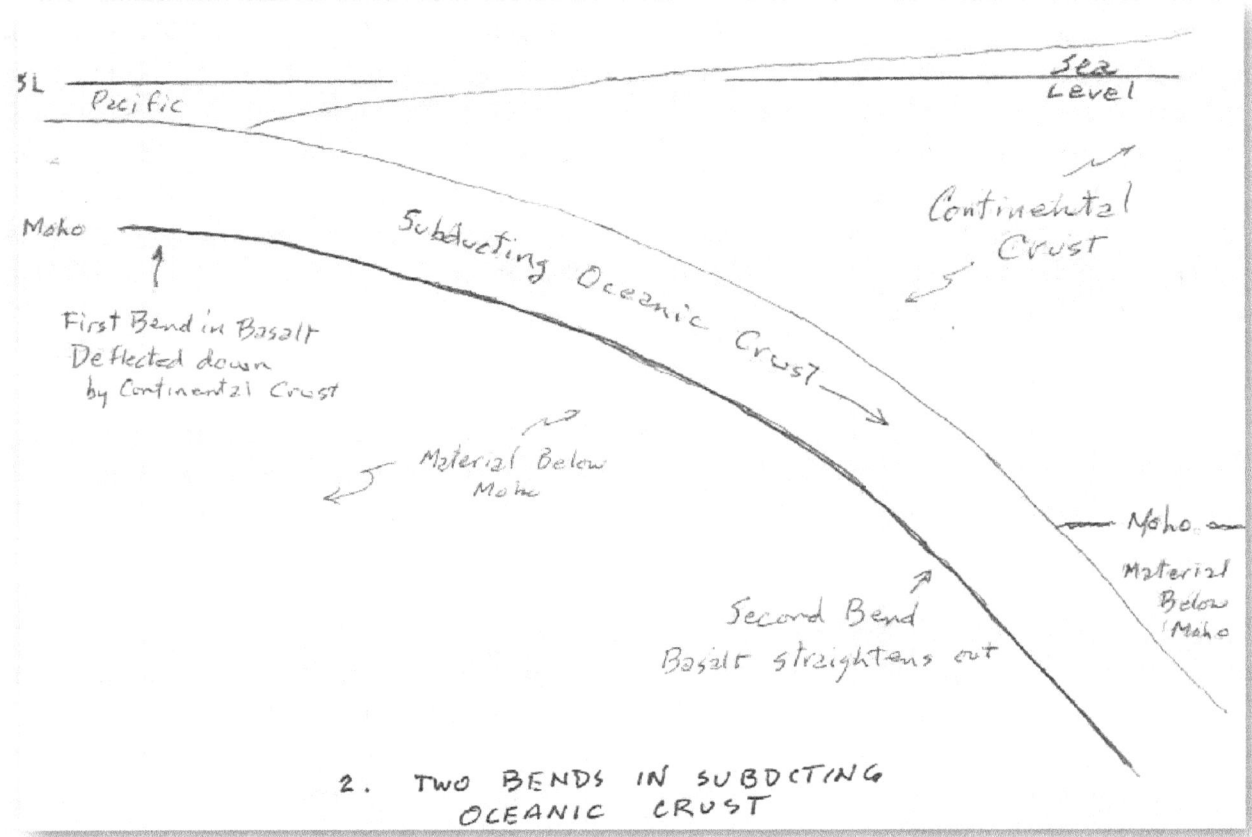

Figure 2. Two bends in a slab of subducting oceanic crust.

Thereafter the slab of oceanic crust slides down at a nearly constant angle. The miles-thick layer of oceanic crust continues to slide down in the channel already established. The basalt continues down in the channel a few hundred miles, before it seemingly becomes assimilated. Below that depth there are no earthquakes, inferring that subducting slabs are no longer brittle. Or no longer exist. When two plates comprising continental crust collide, one continental plate rides on top of the other, and mountains with high plateaus are formed (Himalayas). When two plates of oceanic basalt collide, similarly one may ride up on top of the other, perhaps forming a submarine plateau and the other plate will be deflected down (Marianas Trench).

Six

The Moho

In 1909 a Croatian seismologist, Andrija Mohorovicic, studied the records of vibrations caused by earthquakes. Sudden elastic ruptures within the earth cause these vibrations, which behave like sound. The first arrival of sound from an earthquake will have taken the fastest route to the seismograph. This "fastest route" will favor material with high velocity. Sound must travel from the higher velocity material below, and then up through the shallow material, which has slower velocity. By plotting the arrival times recorded at several seismographs at varying distances from the earthquake, the velocity of sound in deeper, faster layers may be calculated. Mohorovicic found a sharp velocity change, with slower velocity above (about 15,500 miles per hour), and faster velocity below (about 17,500 miles per hour). The Mohorovicic discontinuity became known to linguistically-lazy geoscientists as "the Moho". It occurs worldwide a few miles below the ocean floor, and considerately deeper under continents.

❖ ❖ ❖

Seven

Current Popular Concepts of Plate Tectonics

Following is a general summary of some of current dogma about plate tectonics. These theories are somewhat contentious, and alternate variations exist.

An earth model now in vogue amongst geoscientists has three components from the solid surface down: lithosphere, asthenosphere and core. The lithosphere is rigid. Lithosphere includes the continental and/or oceanic crust and some of the underlying mantle. The mantle is crystalline rock composed of dense, dark, crystalline minerals. At some indeterminate depth, the rigid lithosphere transitions to lower mantle, or mushy asthenosphere, which also comprises dense, dark, crystalline minerals. The asthenosphere is not rigid, so it moves outward from the deep hot root below a mid-oceanic ridge. Below the asthenosphere (mantle) the core is composed of liquid (molten) iron and nickel, with enough fissionable material to maintain high temperature. The inner part of the core is a ball of solid iron/nickel.

In 1929 Arthur Holmes proposed that convection of rising warmer mantle, moving up from the core, spread out under the continents, dragging the continents along. The convection cell was completed by downward convection of the cooled (denser) mantle.

By 2014, with all the geophysical data which has become available, this 'treadmill" theory has not changed much. Movement of the plates is caused by convection within the mantle. When plates collide, the cold lithosphere in the under-riding plate is denser, and it sinks, causing "slab pull", the most important force involved in moving the plates. Gravity sliding off the elevated buoyant mantle under the ridge is called "ridge push". Since the descending slab does not appear to be accelerating, it is concluded that a frictional force of "slab resistance" counters the force of "slab pull". Continental crust has deep roots, and as the continent moves, dragging these deep roots through the asthenosphere creates "continental drag".

"Slab suction" is the force that the subducting plate exerts on the overriding plate. The overriding plate tends to be sucked down by the movement below, creating a back-arc basin in the overriding plate.

Each mid-ocean ridge is like a mountain chain under the sea. The conventional understanding is that the higher elevation of the ridge is caused by the lighter density of the rising mantle plume under the ridge, compared to the surrounding colder and denser mantle. The rising mantle plume has been heated by proximity to the hot core. The mantle plume of buoyant rock is less dense than the surrounding rock, so the plume rises. When the rising plume gets towards the base of the oceanic crust, it splits in two with each side moving away in opposite directions, paralleling the direction of the motion of each plate. As the plume circulates in both directions away from the spreading center it cools, becoming denser, and therefore sinks back down toward the core.

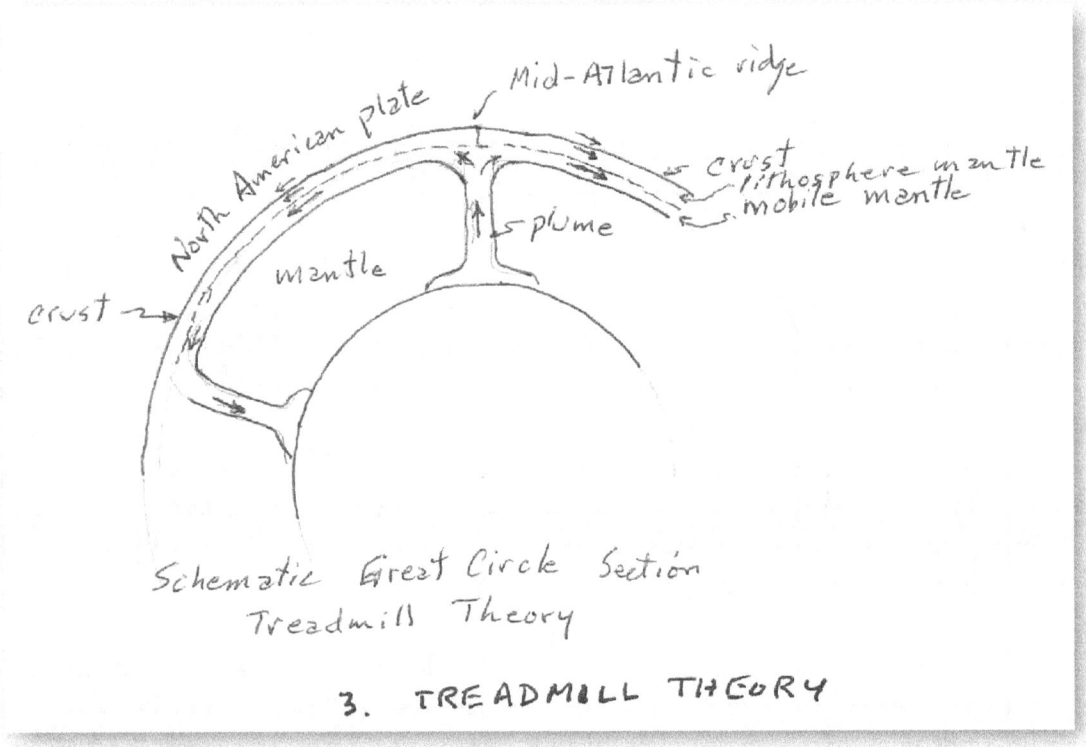

Figure 3. Schematic cross section of the earth showing treadmill theory

Centered below the mid-Atlantic ridge is an en echelon series of parallel lava chambers. These chambers are charged by plumes of molten lava circulating up from the core, or by hot mantle transforming into magma. (Note: **magma** is subsurface molten rock, whereas **lava** is molten rock which is in the process of erupting, either above or below sea level.)) The magma chambers simultaneously feed basalt eruptions along each sector of the spreading center. Each lava chamber feeds an eruption at the same rate as in the adjacent sector. Thus the plate moves as a unit away from the spreading centers.

The rigid lithosphere, including oceanic crust and upper mantle (together about 200 miles thick), is carried along by faster moving mantle on which the lithosphere rests. The two plates are carried from the mid-Atlantic like opposing conveyor belts.

At the distant edge of the moving plates the lithosphere becomes denser as it cools. At a distant edge of a plate there is a collision with another plate. Cooled oceanic crust, together with the underlying rigid upper mantle, are deflected down and slide under the advancing continental crust. The cold, dense slab of oceanic basalt, along with the remainder of the attached mantle, sink down to the core, where the mantle is reheated.

❖   ❖   ❖

Eight

Questioning Current Theory of Plate Formation

A. IMPROBABLE VOLCANIC CHOREOGRAPHY

It seems improbable that in the mid-Atlantic where there are many separate lava chambers, 1) each is linear, 2) these linear segments are parallel, and 3) the volcanic eruptions are coordinated, you might say choreographed, so that each eruption stays in step not only with its neighbor, but also with all the other segments of spreading along the mid-Atlantic ridge. Lava introduced into each sector of the string of spreading centers (separated by transform faults) is just enough to fill in between the two separating plates. In "gasoline-station parlance", the lava filling in between departing plates remains "topped up".

Volcanic eruptions on land are usually messy affairs because liquid lava, being less dense than the surrounding solid volcanic rock, rises buoyantly to follow the path of least resistance. In contrast, the Atlantic spreading centers seem well organized and have provided a continuing source of new basalt at a relatively steady rate for 150+ million years.

B. SUBDUCTING CRUST SHOWS NO SIGN OF "WIND" FROM MOVING MANTLE

Current theory proposes that the lithosphere, 200 miles thick, is dragged along by the moving mantle below. The moving mantle is envisioned as 200 miles thick. So the North American plate, passive mantle above and the faster driving mantle below, together 400 mile thick, moves west. This formidable "wind" should seriously impact the subducting oceanic crust, coming from the west towards the "wind". But it doesn't. According to the shape of the Benioff zones, the subducting oceanic crust around the east side of the Pacific enjoys its freedom of movement and straightens out as soon as it clears the obstructing continental crust of the confronting plate.

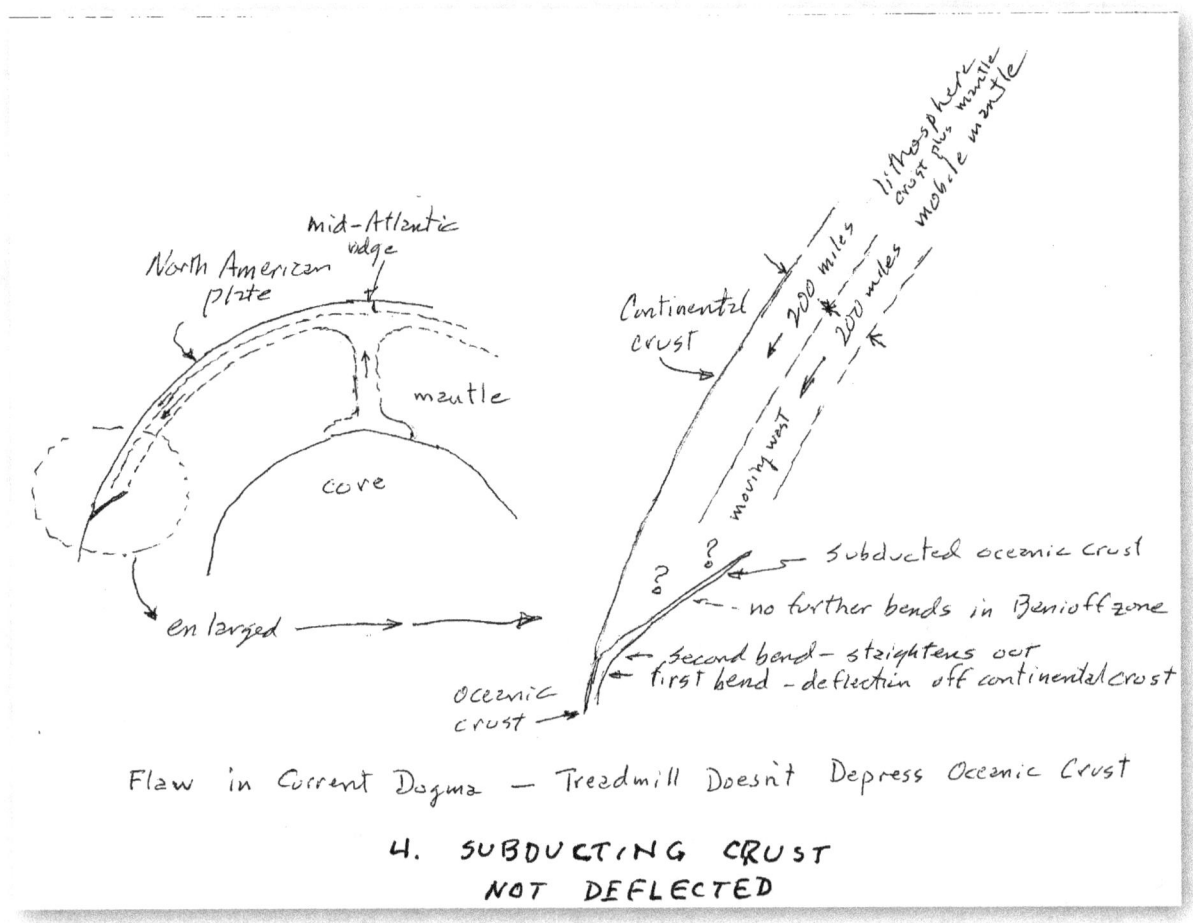

Figure 4. Treadmill inconsistency: moving lithosphere does not deflect subducting slab of oceanic crust

This is a major defect in current "treadmill" dogma.

C. WHY IS THE RATE OF FORMATION CONSTANT?

The thickness of oceanic crust under the abyssal plains of the Atlantic seems reasonably constant over millions of square miles. This implies that this oceanic crust was formed at a constant rate for more than 150 million years. What kept the rate of formation constant? Plumes of hot mantle rising about 1600 miles from the earth's core would not have an automatic control system for maintaining a constant rate of formation of lava. This is another weak spot in the current theory of crustal spreading.

D. WHY IS LAVA FORMATION REGIMENTED?

Lava is capricious. Being less dense, it seeks the ways of least resistance to its upward movement. Below the northern Atlantic, eruption of basalt creates oceanic crust. In addition there are a few submarine seamounts, the Azores and the volcanism onshore Iceland. Harry Hess found extinct volcanos reaching up from the ocean bottom with tops originally eroded flat by wave action. These guyots, as he called them, are covered by progressively deeper water further west and east from the mid- Atlantic spreading centers.

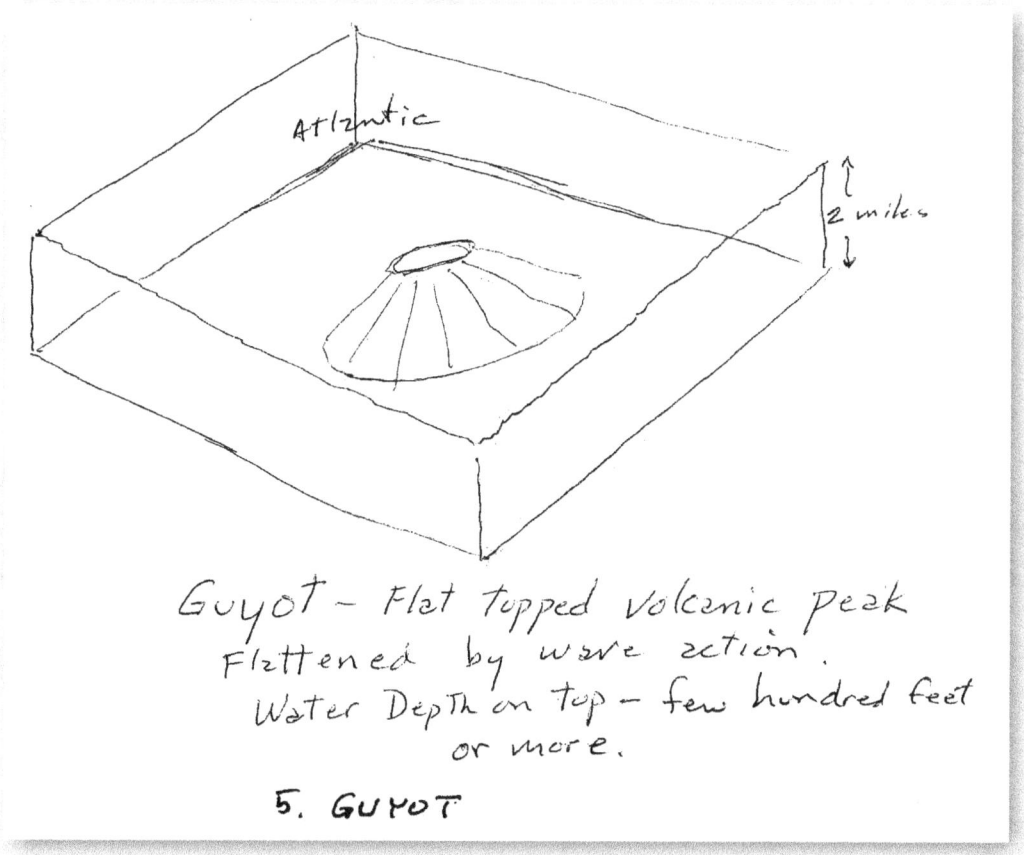

Figure 5. Guyot: submarine flat topped volcanic cone, eroded by wave action.

But compared to the several-mile thickness of oceanic crust under the Atlantic, the volume of these anomalies is very small. It seems that almost all of the quixotic lava ends up as the relatively tidy mid-oceanic basalt from spreading centers.

Rather than liquid lava somehow converted from rising mantle, what alternate mechanism could be responsible for the steady formation of lava at spreading centers? The Moho separates materials with different velocities of sound. What characteristics of the material above and below the Moho would be required to facilitate the generation of oceanic basalt at spreading centers?

Even greater problems for plumes of heated mantle rising from the core are discussed in Chapter 20, **As the World Goes Round**

❖　❖　❖

Nine

The Facts at our
Disposal

There are only about three dozen facts known about the Moho under the Atlantic, and what lies above and below it. The most important are summarized as follows.

A. EARTHQUAKE RECORDS

From the many seismic listening stations around the world, the location and depth of the ruptures in brittle rock (earthquakes) are triangulated using the times of the first arrival of sound and the speed of the sound. The epicenter is the place on the earth's surface directly above the rupture, and the hypocenter is the actual location below the surface where the rupture occurred. Earthquakes occur in either continental crust or oceanic crust. There appear to be no earthquakes generated at the Moho or below the Moho, except in subduction zones. **Conclusion:** the material below the Moho has no strength, which means stress cannot accumulate, and therefore no earthquakes can occur there.

The United States Geological Survey in 1973 started a compilation of all earthquakes recorded world-wide. This record is accessible by you online, but there are so many earthquakes recorded it is difficult to make sense out of it. The Survey solved the problem of overabundance of data by allowing the user to select quakes pertinent to his study. Want a list of earthquakes of magnitude 7.0 (Richter scale) or greater since 1973 around Chile? Just query the USGS website.

Around the edge of the shrinking Pacific there are descending curtains of earthquakes extending hundreds of miles below the adjoining continental and oceanic crust (Benioff zones). Each of these curtains of earthquakes apparently tracks the course of a descending slab of brittle oceanic crust, deflected down after colliding with adjoining continental or oceanic crust. This process is called "subduction". Earthquakes occur when the plate is stressed beyond the breaking point. The slabs of oceanic crust housing these quakes extend down about 450 miles below sea level, encased in the viscous material below the Moho. Eventually the slab of descending oceanic crust apparently changes character, because below that depth there are no more earthquake hypocenters, signaling that the descending slab is no longer brittle.

Assumption: the Moho coincides with the base of the earth's crust. Oceanic crust is a few miles thick. Under continental crust the Moho is much deeper, perhaps 10 to 30 miles. The United States Geological Survey has compiled a world-wide map of the depth to the Moho.

This contour map of the thickness of the Earth's crust was developed from the CRUST 5.1 model. The contour interval is 10 km; we also include the 45 km contour for greater detail on the continents.

6. USGS MOHO CONTOURS

Figure 6. 1998 USGS map of Moho depth

B. SEISMIC SURVEYS

Earthquakes generate two main types of sound waves transmitted through the earth and recorded on seismographs. The faster sound waves are pressure pulses, or "P" waves, and the transverse "S" waves are much slower. A sideways jerk to the end of a stretched rope will cause a bump to move down the rope (S wave). In air, water or other liquids, the S wave cannot be transmitted. In solid materials both P and S waves are transmitted. The material below the Moho transmits both P and S waves, so it acts like a solid, on a milli-second time scale, during the passage of sound waves.

At a given depth, the speed of P waves is about twice that of the S waves.

Seismic investigations use naturally occurring earthquakes or man-made sources of energy (dynamite, nuclear tests, etc.). Devices listening for earthquakes are distributed almost exclusively over the continents, and they are operated 24/7. These seismometers measure the arrival times and amplitudes of earthquake vibrations. Sound arriving first to the listening device has taken the fastest route, which depends on the velocities of sound in the different layers. Analyzing these first arrivals is **refraction** seismic. The Moho was first detected using refraction seismic.

Commercial seismic crews, working for oil companies, measure **reflection** data. They create a noise near the surface and record the resulting echoes from underground reflectors. A continuous reflection is correlated from seismic line to line, and a contour map is constructed. This map of reflection times is converted to depth by using the velocity of sound. The velocity may be calculated, or better yet, measured in nearby wells. These maps are used to evaluate the likelihood of finding petroleum with the drill bit. Commercial crews cease recording the reflections (echoes) after a few seconds, so these surveys typically gather information down to a depth of only a few miles below the surface.

Occasionally a reflection seismic crew will record for tens of seconds, long enough to "hear" the echo from the Moho. Whereas the Moho is always observed world-wide using **refraction** seismic, only sometimes is there a recognizable **reflection** (echo) from the Moho. The velocity contrast across the Moho is about 12 %, which should be enough for a good reflection. But the recording equipment and the earth's response must work in harmony. Perhaps the Moho reflection in some areas is weak, or perhaps it is masked by noise and questionable signals. At depths of 10 to 30 miles below the surface to that reflector, it is not unexpected that seismic systems don't always produce the fidelity sought.

The thickness of oceanic basalt is analogous to the skin of an apple – very thin compared to the diameter of the earth. How thin? One way to measure the thickness accurately would require good seismic reflections from both top and bottom of mid-oceanic basalt. Also needed would be accurate velocity of sound in that interval. But reflection seismic in volcanic rocks is notoriously poor. Sound bounces around off fractures and seldom can be resolved to an unambiguous solution. Perhaps the best means of estimating the thickness of oceanic crust would be a refraction seismic survey with both sensors and sources of sound placed on the sea bottom.

C. LITHOSTATIC PRESSURE BELOW MOHO

In oil wells, which may reach depths of several miles, there are fluids in the porosity of the rock (primarily water, oil and gas). Pressures in these fluids often are not hydrostatic, i.e. sometimes pressures are less than a column of water from the surface to that depth, and sometimes pressures are greater. High pressures in the water in those pores may approach the pressure of the combined weight of the rock and the fluids at that depth (lithostatic pressure). In such high-pressure zones the strength of the rock decreases to a negligible level because the fractured rock frame "floats" in the fluid. By analogy, the **assumption** is that since the solid material below the Moho never fractures (no earthquakes), it behaves as if its internal pressure is lithostatic.

D. EARTH'S CORE
The earth has a dense core (about 4,300 miles in diameter) of iron-nickel, as deduced from the strength of the earth's gravity. The earth's core does not transmit S (shear) waves originating on the other side of the globe, so it is deduced that the core is liquid. Only the faster P (pressure) waves transverse the core. On closer inspection of seismic recordings, in 1936 a Danish geophysicist, Inge Lehmann, determined that the earth's liquid core has a solid inner core.

The age of the oceanic basalt underlying the earth's oceans is no older than about 230 million years, whereas the earth is about 4.6 billion years old. The Atlantic Ocean has been spreading for 150+ million years. The age of spreading is known from the ages of sediments in offshore oil wells, drilled above the oldest basalt.

E. EARTH MAGNETISM
The earth's magnetic poles migrate. Old topographic maps published by the United States Geological Survey show not only the difference between true north and magnetic north, but also the annual rate of change.

After lava erupts in a spreading center, it cools and solidifies to basalt at about 1400 degrees Fahrenheit. When the solid basalt cools just below the Curie point (about 1060 degrees Fahrenheit), the orientation of the earth's magnetic field at that time is imposed on the crystals of magnetic minerals in the basalt. That magnetic orientation remains fixed unless the material is reheated above the Curie point.

In 1963 Frederick Vine, Mathew Drumond and Lawrence Morley recognized that magnetic stripes in the submarine oceanic crust were records of reversals in the polarity of the earth's magnetic field. The Pacific is older than the Atlantic. For the past 230 million years under the Pacific these reversals occurred sporadically, on the order of every million years, except for one period of several million years with no reversals. Why, how and when does the earth's magnetic field reverse? As yet there is no obvious answer.

F. SUBMARINE TOPOGRAPHY
The principal reason the earth's ocean bottoms were mapped was that major navies wanted to know where they could hide their submarines. As time went by, Google Earth were able to obtain the data and make it available on the internet.

The submarine topography of the mid-Atlantic ridge varies greatly. In some places there is a north-south cleft observed on the sea bottom, giving evidence of the line of separation between the two departing plates. Elsewhere along the mid-Atlantic ridge, the ocean depths given on Google's oceanographic map show no significant topographic demarcation. In fact, there are hills a couple thousand feet high, here and there, along the ridge. Apparently lava spilled out onto the ocean floor and built up as lop-sided hillocks. These hills sometimes cover the line of separation between departing plates.

The crest of the range averages about two miles below sea level, whereas on the adjacent abyssal plains the average depth of the ocean exceeds three miles. Occasional guyots crest at increasing water depth away from the mid-Atlantic ridge.

There are also a few submarine volcanic peaks, "seamounts", on the ridge. One volcanic island chain (Azores) straddles the mid-Atlantic ridge.

Sparker profiles are inexpensive seismic surveys, but their main drawback is that the results are shallow. The few sparker lines in the Atlantic, found by your author in geologic literature, depict thin "ocean dust" deposits, on top of the hummocky basalt. This oceanic dust deposit thickens very gradually going away from the Atlantic ridge. The significant result is that the ocean dust is not shown to be offset by faults. If the dust layer is not offset by faults, then it can be safely concluded that the underlying basalt has not been faulted in the time since the dust settled on the basalt.

Closer to shore are thick sediments eroded off the adjacent continent. These sands, shales and carbonates are carried by rivers into the sea, forming coastal sediment systems, continental slopes and distribution systems for sediments on the ocean floor. Here is the hunting ground for oil and gas on both sides of the Atlantic.

The chemical content of new basalt from the mid-Atlantic ridge and from the oceanic crust in the Pacific is remarkably similar, as determined by oceanographic researchers. They analyze grab samples from the ocean floor, and from cores drilled into the ocean bottom.

G. TOMOGRAPHY IS WHAT?

Tomography is one of the whizz-bang new technologies generously utilizing powerful computers. Medical tomography is big business. Seismic tomography uses an algorithm in software designed to reconstruct and analyze tomographically the seismic energy (both P and S waves). The object of the tomographic processing is to get a three dimensional display of the velocity of sound all the way down to the earth's core.

This velocity derived from tomography is interpreted to be a proxy for temperature. Results were a bit hazy as of 2010. More seismographs are being installed to increase definition. Researchers have been trying to locate the postulated plumes of hot material rising from the earth's core under spreading centers. As of 2010, no plumes had been detected. This suggests that rising plumes leave no measurable trace, or are non-existent.

H. OPHIOLITES

Fault zones in some mountain ranges around the world have remnants of ancient oceanic crust, called ophiolites. These ancient remnants are identified as oceanic basalt by the intimately overlying dusting of minute fossils of floating organisms ("ocean dust") which rained down in the ancient ocean on top of the oceanic basalt, just like what happens today. Ancient oceanic sea bottoms have been caught up in mountain building, and squished in fault zones. These ancient mid-oceanic basalts, called "peridotite" show that mid-oceanic spreading centers existed hundreds of millions of years before today's volcanic activity under current ocean ridges. Peridotites are exposed in outcrops on land where they may be studied, which is much handier than trying to study oceanic basalt under two or three miles of water. Peridotites apparently endured compression during their tumultuous geologic history, for their density averages 3.3, whereas modern ocean floor basalt averages 2.9.

I. SATELLITES

Satellite measurements of the earth's gravity field yield estimated depths to the Moho. The European Space Agency and Politecnico di Milan fund this GEMMA project which uses an algorithm to process gravimetric data. Results vary according to individual interpretation

Ten

Late breaking news!

"Late breaking", in this case is not on the time scale of minutes or hours, common to modern media. **Moho Motion** has been coming together at a speed which would qualify snails for speeding tickets. Write and rethink. "Honey do's. Prune fig tree. Serious attack of laziness. Again write and rethink. Blind alley. Procrastination rules! The Rotary Club of Bakersfield. Etc.

Furthermore, the author's tracking of developments in the geologic literature has been spotty, given limited access to the massive volume of current geoscience publication, and given the fact that very little thereof pertains to the new concepts presented here in **Moho Motion.**

A. PROBLEM FOR MOHO MOTION

Here is the fly in the ointment. The June 8, 2007 issue of Science has an article by M.R. Brudzinski *et al* entitled "Global Prevalence of Double Benioff Zones". They studied the curtains of hypocenters of earthquakes around the Pacific where subduction is taking place. Their conclusion is that most Benioff zones are really not just one descending curtain of earthquakes, but rather **two** curtains, each about 6 miles thick, separated by a 12 mile thick dead zone without earthquakes. Furthermore, they mention that the concept of double Benioff zones has been around for thirty years (unnoticed by your author).

The theory postulated in this **Moho Motion** is that one Benioff zone, a few miles thick, is the source of earthquakes in the descending slab of brittle oceanic crust around the Pacific. The concept of double Benioff zones does not fit with the concept presented in this work. Can this apparent contradiction be squared?

Let's consider dimensions. The sloping Benioff zones accompanying thousands of miles of seafloor trenches, cover very large areas around the Pacific down to depths as much as 450 miles below sea level. Triangulation with sound waves to determine an accurate location of a hypocenter assumes precise knowledge of the three-dimensional distribution of velocity of the P wave, and the exact time of the first arrival of sound at the seismograph (how good is the "pick"?). P wave sound travels at Benioff depth about 17,500 miles per hour, so a second here or there can make a significant difference in the computed location of a hypocenter. Next, there is the difficulty of projecting these locations onto a vertical plane crossing the Benioff zone at a right angle. The swath of projected hypocenters may be many tens of miles wide in order to get a meaningful number of data points. So, how good are the projections?

Brudzinski *et al* have answers for these potential shortcomings. In northeastern Japan there is a relatively dense array of seismographs, and earthquakes are relatively frequent. Therefore the data are very good.

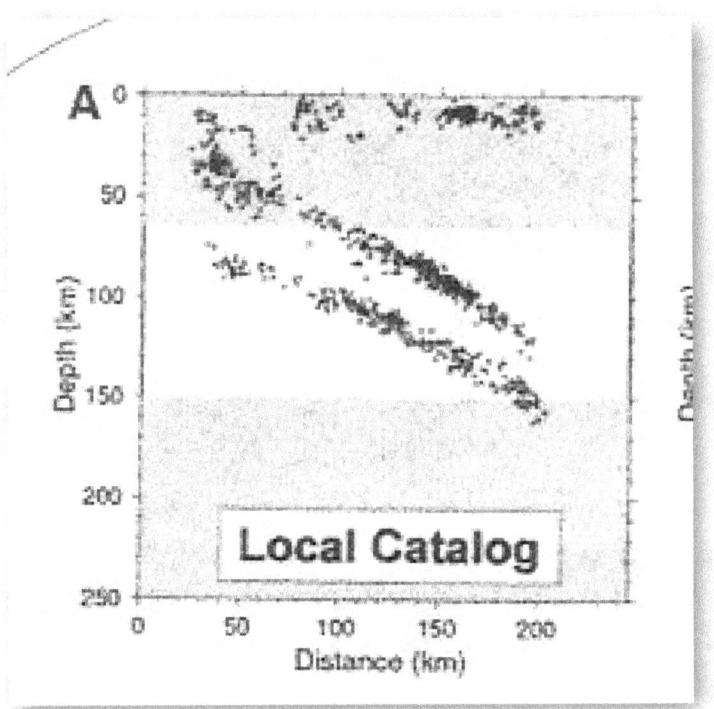

Figure 7. Display A, Local Catalog. From M. R. Brudzinski, C. H. Thurber, B. R. Hacker, and E. R. Engdahl: **Global Prevalence of Double Benioff Zones**, in Science 8 June 2007, Vol. 316, no. 5836, pp 1472-1474. Reprinted with permission of AAAS. This display charts earthquake hypocenters under northern Japan. A double curtain of earthquake hypocenters from subducting oceanic basalt are projected onto the plane shown. The Benioff zone is seen from the north, so west is to the right.

The double Benioff zones in Brudzinski's illustration appear to be definite, clearly defined and parallel.

B. A POSSIBLE EXPLANATION

An acoustic trick may create the second, lower layer of phantom hypocenters in figure 7.

Sound is reflected from surfaces separating materials with substantially different velocity. The echo of your voice from a wall of granite (Yosemite) is a good example. The Moho separates materials with about 12% difference in velocity, so it should be a good reflector. Both top and bottom surfaces of a subducting slab of oceanic crust are Moho surfaces.

In Brudzinski's Figure 7 the upper band of hypocenters extends all the way up from 130 kilometers subsea to a depth of about 40 kilometers. This is the logical depth for the base of the continental crust. This is interpreted to mean that the upper band of hypocenters occurred in the uppermost part of the subducting slab of oceanic crust.

Computers select first arrivals of sound. The expanding bubble of sound from an earthquake in the subducting slab under Japan will take two routes to the earth's surface. The first will go directly up. A second route will go down to the bottom of the slab and be reflected back up.

Good geological thinking here merges these observations into a highly probable conclusion. The upper band of hypocenters is from the uppermost part of the subducting crust. That is where the earthquakes occur. The parallel configuration of the two bands, which even track the bending in the slab, indicate that the lower band is a reflection, and the likely reflector is the Moho surface at the bottom of the subducting slab. The dead zone in Figure 7, between slabs, represents two-way time for sound to round trip from top to bottom in the subducting slab.

Checking whether this possible phantom layer exists would be easy. Identify each hypocenter in the shallow layer in Brudzinski's Figure 7 and discard any reflections detected up to 10 seconds later. Replot the surviving hypocenters, as in figure 7, and see if the lower group of hypocenters has thinned or disappeared. "Disappeared" equals "phantom".

The rejected "hypocenters" (the phantoms) could be used to estimate the thickness of the oceanic slab there. Multiply half the time-difference by velocity in the slab, to get thickness.

For the rest of the story explaining the cause of these earthquakes, please refer to the last section in Chapter 20, **As the World Goes Round.**

❖ ❖ ❖

Eleven

Devil's Advocates Find Inconvenient Facts

In the seventies when plate tectonics came into vogue, devils advocates like Art Meyerhoff pointed out that the new theory had serious shortcomings. Most glaring was Iceland. According to the newfangled idea of plate tectonics, Iceland, being a volcanic island smack dab on the mid-Atlantic spreading center, "should" split apart, and each half should be attached to a departing plate. But Art pointed out that there were fossils on Iceland more than 10 million years old, and this proves that Iceland was not drifting apart, even though it is located in the middle of the mid-Atlantic ridge.

More recently, in 2000, David Pratt emphasized that many rocks collected by dredging and shallow core holes in the ocean bottom on the mid-Atlantic ridge are the wrong type and hundreds of millions of years too old to fit the theory of plate tectonics (i.e., rocks at the spreading center should be exclusively basalt, and the basalt should be youngest at the spreading center). Radiometric dating of rocks is fairly accurate, so Pratt's objection must be addressed.

Twelve

Other Facts worth Noting

Keep in mind the laws of physics. Not Einstein's, but Newton's. Example: conservation of momentum, whether in a straight line or in the revolving earth. And harken to the discipline of mechanical engineering, especially the study of strength of materials. Poisson's Ratio, buoyancy, soil mechanics and common sense need to be addressed. There is a significant advantage in taking a multi-disciplinary approach in devising a new theory.

Geologists are uniquely suited to assemble and diagnose incomplete disparate facts and ideas. The stock in trade of engineers, chemists, mathematicians and physicists is an exact answer to a physical problem. Geologists are satisfied finding a solution which seems to fit well. Can you imagine being a member of an exploration team, proposing several multi-million dollar projects, and being euphoric when just one discovery promises to be a commercial success? Such is the happy lot of the successful petroleum explorationist.

❖   ❖   ❖

Thirteen

Putting the Facts Together (House of Cards??)

Geologists are pragmatic. In trying to make sense of the geology of rocks below the surface of the earth, a geologist will examine the few patches of exposed rock in the area, and reflect on similar geology seen elsewhere. Fitting the meager facts into his catalog of geologic experience, a geologist will construct a plausible four dimensional) model (time is the fourth dimension). The model will stand until someone else has a better idea, or until better data require changes.

In that spirit, let's build a house of cards representing the mid-Atlantic ridge, using observed facts, but interpreted in a different way. First let's start with a concept. Then as we go along let's review the data, consider the current theory, introduce fresh assumptions, arrive at conclusions and see whether the concept seems to fit the data.

Let the procedure begin and see where it leads. Maybe this will **not** be a house of cards?

A. CRUST FLOATS ON MOHO

It is **assumed** that the crust, whether oceanic or continental, floats on the material immediately below the Moho. The crust is decoupled from the material below the Moho. The Moho is the shallowest world-wide velocity anomaly, and it has been extensively recognized and to some extent mapped. Current dogma also places the base of the crust at the Moho. So, let's go with the Moho as the separating surface between crust and the material below, and see where this leads.

Current dogma has lithosphere being dragged along as a separate entity by the moving mantle below. The concept is that the base of the earth's rigid plates is about 200 miles below the surface of the earth. Below that, another 200 miles down is the base of the moving mantle. This moving mantle drags the overlying plate along, sort of like a conveyor belt.

In contrast, the proposed theory in **Moho Motion** raises the surface of movement of the crust up almost 200 miles to the Moho. The Moho is an interface with the crust above floating on the "liquid-like" material below. There is no friction between the crust and the material below the Moho, because that material has no strength (evidenced by no earthquakes) so it cannot resist movement.

As an eager engineering freshman will triumphantly explain, if a fly lands on a railroad rail, the rail must either bend or break. He is right of course, and the insight has punch because of the ridiculous, impractical scale. But in the case of crust floating on the Moho, there is no bending or breaking. Floating is floating.

B. ASSUME LAVA SOURCED AT MOHO

In conjuring up a new explanation of the mechanics of seafloor spreading, we start by acknowledging that many tens of millions of cubic miles of basalt have been generated under the mid-Atlantic ridge. That's too great a volume for tenuous long distance transport. There must be a simpler way. Rather than hot mantle coming up over a thousand miles from the core and somehow transforming to lava, let's **assume** that the lava is sourced at a spreading center. But what mechanism could cause lava to form there?

Could the material below the Moho melt whenever the confining pressure drops below its stability condition? Normally the material below the Moho is at lithostatic pressure. What if the pressure on **top** of the Moho is reduced below the otherwise ubiquitous lithostatic pressure? Then in local spots **below** the Moho, could the material just below the Moho become unstable and melt? Is mid-Atlantic basalt generated at each spreading center by local reduction of pressure on top of the material below the Moho? If so, this implies that the material below the Moho is stable only as long as lithostatic pressure prevails. The material below the Moho can exist only within its stability envelope, where pressure is lithostatic, or nearly so.

Assumption: if the confining pressure on the material below the Moho is reduced below the stability threshold, the material will spontaneously change into molten lava.

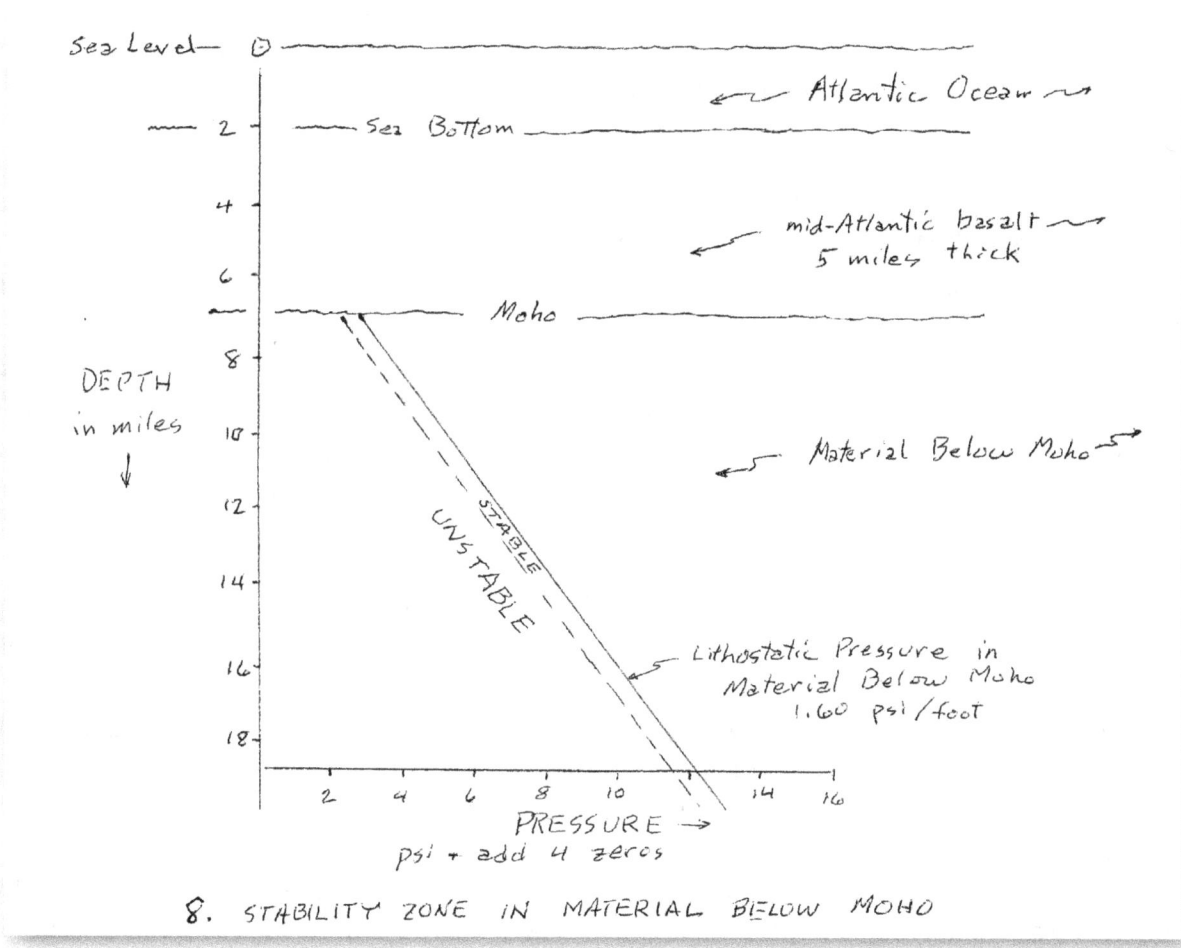

Figure 8 Below the Moho the material is stable at lithostatic pressure. If the pressure just below the Moho is reduced enough, that material will become lava.

C. MAKEUP OF MATERIAL BELOW THE MOHO

The makeup of the material below the Moho must include a continuous mat of crystals in order that seismic sound waves are transmitted as if it were a solid. Crystals in the mat may be the same minerals found in oceanic basalt. Below the Moho the material must be at lithostatic pressure (no earthquakes). Thus the mat must be floating in a viscous fluid at lithostatic pressure, possibly a "soup" of high-energy ions, facilitating recrystallization and eliminating strength in the mat. The lithostatic "soup" floats the mat. Assuming lithostatic pressure logically implies that the dense material below the Moho is mushy.

The change of velocity of sound across the Moho results from a physical difference between oceanic basalt and the mushy material below the Moho. Collected samples prove that oceanic basalt is a hard, elastic rock. On the other hand, the material below the Moho is denser, based on the increased velocity there. The lack of any buoyant action below the Moho is a further indication that the density below is greater than the density of the basalt above. The mush would readily act buoyantly (like salt) if it were less dense than the overlying oceanic basalt. Buoyant displacement would be seen in reflection seismic data.

D. ASSUME BASALT IS IN A DIFFERENT STATE THAN THE MATERIAL BELOW THE MOHO

There has been a bodacious amount of oceanic crust generated under the mid-Atlantic ridge during the past hundred plus million years. Where could it come from?? As long as we are building a house of cards, let's **assume** oceanic crust and the material below the Moho have the same chemical composition, but they are in two different states. Lava is created by some mechanism which transforms the material below the Moho into lava, which then rises and solidifies, forming mid-Atlantic basalt. But how would this transformation occur?? How can material with one chemical content have two different densities? Patience. A case is being built.

E. MOHO MILE HIGHER UNDER MID-ATLANTIC RIDGE

Why is there a ridge under the Atlantic? For an answer we examine the submarine mountain range. The range is about a hundred miles wide, and its crest is more than a mile higher than the abyssal plains on either side. We have noted that the seeming continuity of the thin layer of dusting indicates that the basalt crust below the dust is also undisturbed. Furthermore, earthquakes are confined to the spreading centers on the mid-Atlantic ridge. Both lines of evidence point to crustal quiescence away from the spreading centers.

By the time that the new basalt on the crest of the ridge has moved away 50 miles, the top of the basalt drops a mile in elevation. The thickness of the oceanic crust under the abyssal plains may be constant. Therefore it is reasonable to assume that the rate of formation of lava at spreading centers probably has been essentially constant since the Atlantic formed. Because the ridge is a mile higher, and the thickness of oceanic crust is constant, we **conclude** that the Moho is a mile higher under the ridge than under the abyssal plains.

F. GRAVITY MOVES PLATES

The material below the Moho has no strength, so it provides no resistance to sliding motion of the overlying crust. The massive basalt crust being formed at a spreading center will slide downhill (east and west) off the high Moho centered under the mid-Atlantic ridge. The energy for moving the two plates away from the ridge is gravity. Gravity causes the two plates to slide downhill, losing one mile of elevation from the ridge top to the abyssal plains on either side.

Only one mile of potential energy seems little to move the North American plate, which is several thousand miles wide. But on either 50 mile-wide flank of the ridge, the thickness of the oceanic crust will exert a formidable force as it loses one mile of elevation, sliding down off the high Moho. Although the North American plate moves less than an inch a year, it has considerable momentum (mass times velocity), because of the enormous mass of the crust comprising the North American plate from the mid-Atlantic to the California coast. This momentum keeps the motion of each plate continuous without fluctuations.

G. MECHANISM CONTROLS HEAD OF MATERIAL BELOW THE MOHO

Gravity sliding off the mid-Atlantic high has been going on since the Atlantic first parted. Therefore some mechanism in the material below the Moho has replenished the volume of that material which was transformed into basalt. This mechanism also kept the Moho a mile higher beneath the ridge for all that time. Note that this undefined mechanism in the material below the Moho has maintained higher "head" (to borrow a term from hydraulics), under the mid-Atlantic ridge, and a lower head under the abyssal plains.

"Head" is a measure of potential energy, or the height to which a fluid would rise in a standpipe. Of course a standpipe would not be suitable for measuring the potential energy of the material below the Moho, but that hurdle can be bridged. In Figure 22 the head of the material below the Moho is graphically charted.

It is not necessary to understand how the undefined mechanism in the material below the Moho adjusts the head of that material. Just recognize that it does.

H. FORMATION OF LAVA

The pressure from sea water plus rock above the Moho is lithostatic pressure at a specific location. At a location on the Moho, lithostatic pressure down just balances the pressure up from the "liquid" material below the Moho. By definition, this is "floating". Movement in the mushy material below the Moho occurs slowly. If the overburden pressure on the Moho is reduced locally an appropriate amount, and the material below the Moho cannot regain equilibrium fast enough at that location, it will commence transforming to lava. Lava will continue flowing until the back pressure from the rising lava brings the pressure on top of the material just below the Moho at that location up to the stability pressure. Then flow ceases.

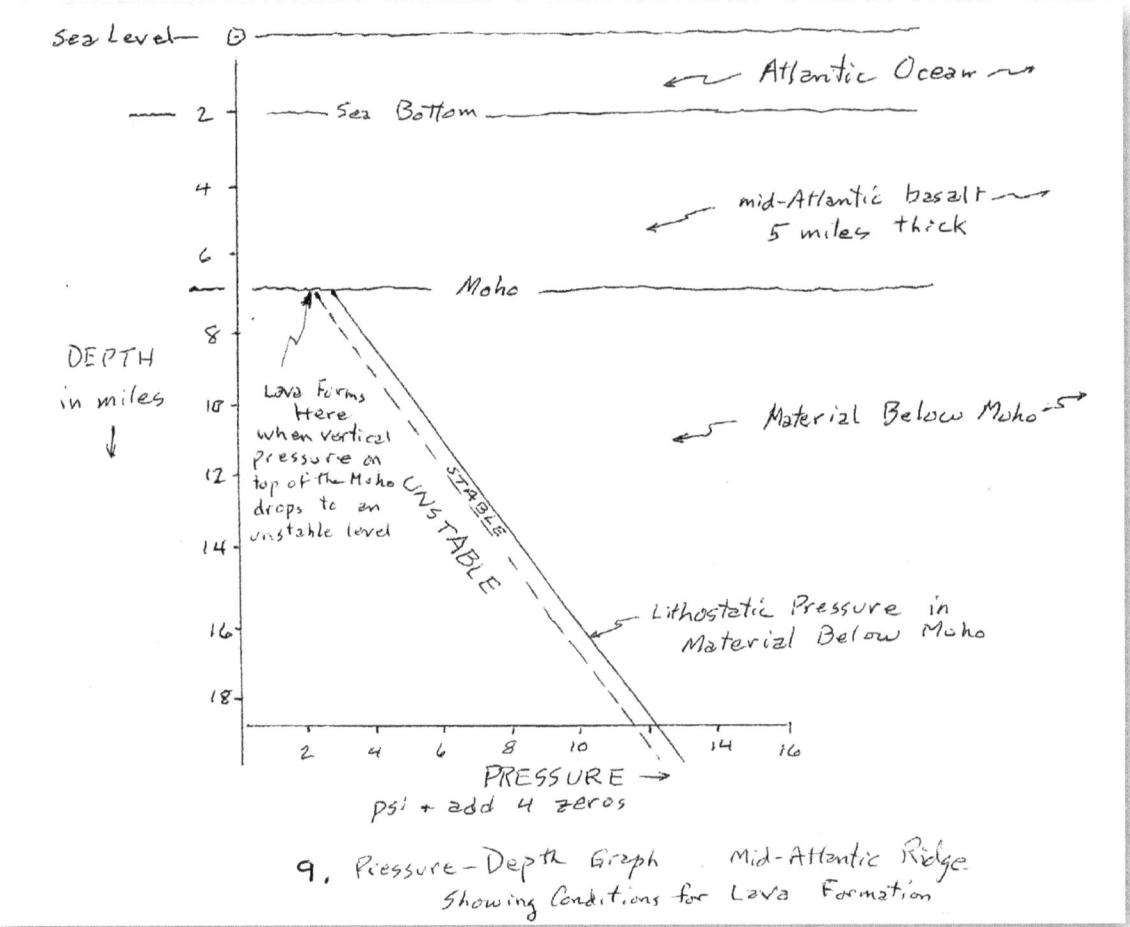

Sea Level — 0

2 — Sea Bottom

Atlantic Ocean

4

mid-Atlantic basalt
5 miles thick

6

Moho

DEPTH
in miles
↓

8

10

Lava Forms
Here
when vertical
pressure on
top of the Moho
drops to an
unstable level

UNSTABLE

STABLE

12

Material Below Moho

14

Lithostatic Pressure in
Material Below Moho

16

18

2   4   6   8   10   14   16
PRESSURE →
psi + add 4 zeros

9. Pressure—Depth Graph   Mid-Atlantic Ridge
Showing Conditions for Lava Formation

Figure 9. Pressure-Depth graph of stable/unstable pressures in material below the Moho under the mid-Atlantic ridge. If the vertical pressure **on** the Moho is less than the stable pressure immediately **below** the Moho, lava forms just under the Moho.

It's sort of like a mechanized water trough on the prairie. Water flows only when cattle quaff!

I. STRESS DISTRIBUTION AT SPREADING CENTER

(Now we come to the weakest level in our house of cards. Skeptics, sharpen your attention! Here we deal with the distribution of stress in the zone of separation between departing plates.)

How is the pressure reduced below lithostatic on top of the Moho (in order to cause lava to flow) when all the overlying rocks and sea water remain in place above it?? The answer lies in the stress distribution within the new oceanic basalt which is being pulled apart at the spreading center. Two phenomena account for this local "levitation".

First, in the basalt the stress of the "pull-apart" causes the basalt to fracture and fault in a predictable fashion. Faults and joints, trending north-south, will dip west and east, 45 degrees or more. North-south oriented irregular horizontal prisms of faulted/jointed basalt are formed in response to the west-east movement in the "pull-apart" zone. Buttressing one another in turn, these prisms locally redistribute vertical load on the Moho. From time to time, when prisms shift because of the ongoing faulting, vertical pressure at a particular location on the Moho will change. Lithostatic pressure directed downwards on the Moho is higher-than-normal in some areas, and lower in others.

Second, in the upper part of the spreading center which is cooled by the sea water, the cold, brittle edges of the departing plates lean against each other, transferring some of the vertical load on the Moho as much as miles east and west. (Good analogies are the natural bridges in the Natural Bridges National Monument in southwest Utah.) As later intrusion of basalt pushes the two sides apart, "leaning" will be reduced locally, causing changes in the vertical stress in the rock frame.

J. PRINCIPAL STRESSES

Stress in deformed elastic solids is measured conventionally in three directions at right angles. These directions are called the principal stresses.

Bruno Sander in 1948 and 1950 published the defining books describing the fabric of rock deformation. He showed that the axis of symmetry of deformation parallels the intermediate stress.

Conversely, if you know the orientation of the axis of bending, that must be the direction of the intermediate principal stress.

Under the mid-Atlantic ridge the two plates slide away east and west in the direction of the least principal stress. The greatest principal stress is the vertical load of basalt and water. That leaves the intermediate stress in north-south direction.

The principal stresses and generalized shape of joints and faults in the spreading zone under the mid-Atlantic ridge are shown in Figure 10.

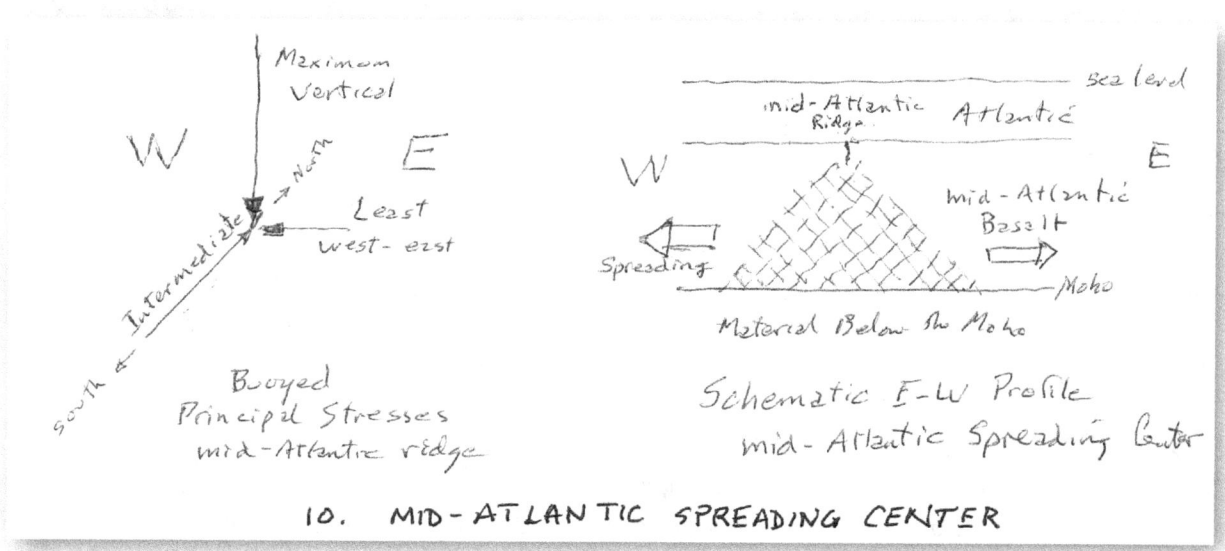

Figure 10. Mid-Atlantic ridge –buoyed principal stresses, and schematic W-E profile

K. LAG TIME

The Moho is an interface separating the crust above floating on the material below. So, if the basalt is floating, why doesn't the pressure immediately below the Moho adjust quickly to pressure imbalances? Apparently there is a lag time. The material below the Mono acts like a solid when sound waves pass through it. But the same material acts like a mush when accommodating an intruding slab of subducting oceanic crust. This suggests that the material below the Moho acts as if it has high viscosity, so it responds very slowly.

L. POISSON'S RATIO

The oceanic basalt rests on the Moho. The mushy material below the Moho exerts no horizontal stress on the oceanic crust. In comparison, the upper cold layer of the crust is brittle and holds each plate together so there can be no horizontal movement within the plate. Therefore, a west-east cross section of a mid-Atlantic spreading zone should look like a fat tepee, coming to a point at the top and spreading at the bottom.

The effect of the rock's Poisson's ratio is to keep rock faces in contact at these great depths because the rock bulges out sideways in response to vertical loading. In the west-east direction of pull-apart, the direction of the least principal stress under the mid-Atlantic ridge, the horizontal compressional stress in the rock frame will be much less than stress in the north-south direction.

M. NO VOIDS, EXCEPT NEAR SEAFLOOR

In the lower part of the oceanic crust under the mid-Atlantic ridge it is unlikely that any cracks will actually open up to create voids. The effects of vertical loading and Poisson's ratio insures that there will be substantial west-east compression, even though the basalt is being pulled apart in each spreading center. Buoyant lava will exploit the reduced west-east compressional stresses to find the easiest egress up the fracture/joint system, starting just below the Moho. Occasional exceptions exist at the sea floor where cold basalt fractures, leaving clefts visible on the sea bottom.

N. ANISOTROPY

Geophysicists will find that the speed of sound in the new oceanic crust below a mid-Atlantic spreading center is faster along the length of the horizontal prisms in the north-south direction, and slower in the west-east direction cutting across all the unhealed fractures and joints. Under the abyssal plains, where the fractures and joints are sealed shut by solidified lava, sound will be faster than in the pull-apart zone. But under the abyssal plains too, sound will be faster in the north-south direction of the welded prisms.

O. BUOYANCY

A further complication is the effect of seawater, which buoys the basalt, reducing the stress in the rock frame. The relationships between head, pressure-depth, and the complications of seawater and Poisson's ratio are a little tricky, so a separate chapter with many diagrams offers further explanation (Chapter 15).

In **Moho Motion**, initials are not used for repeating names. Don't want to scare off the general public in a blizzard of ABCs, requiring back-paging to find definitions. But here is an exception: **psi, for pounds per square inch.** "Pounds per square inch" would take too much space on figures.

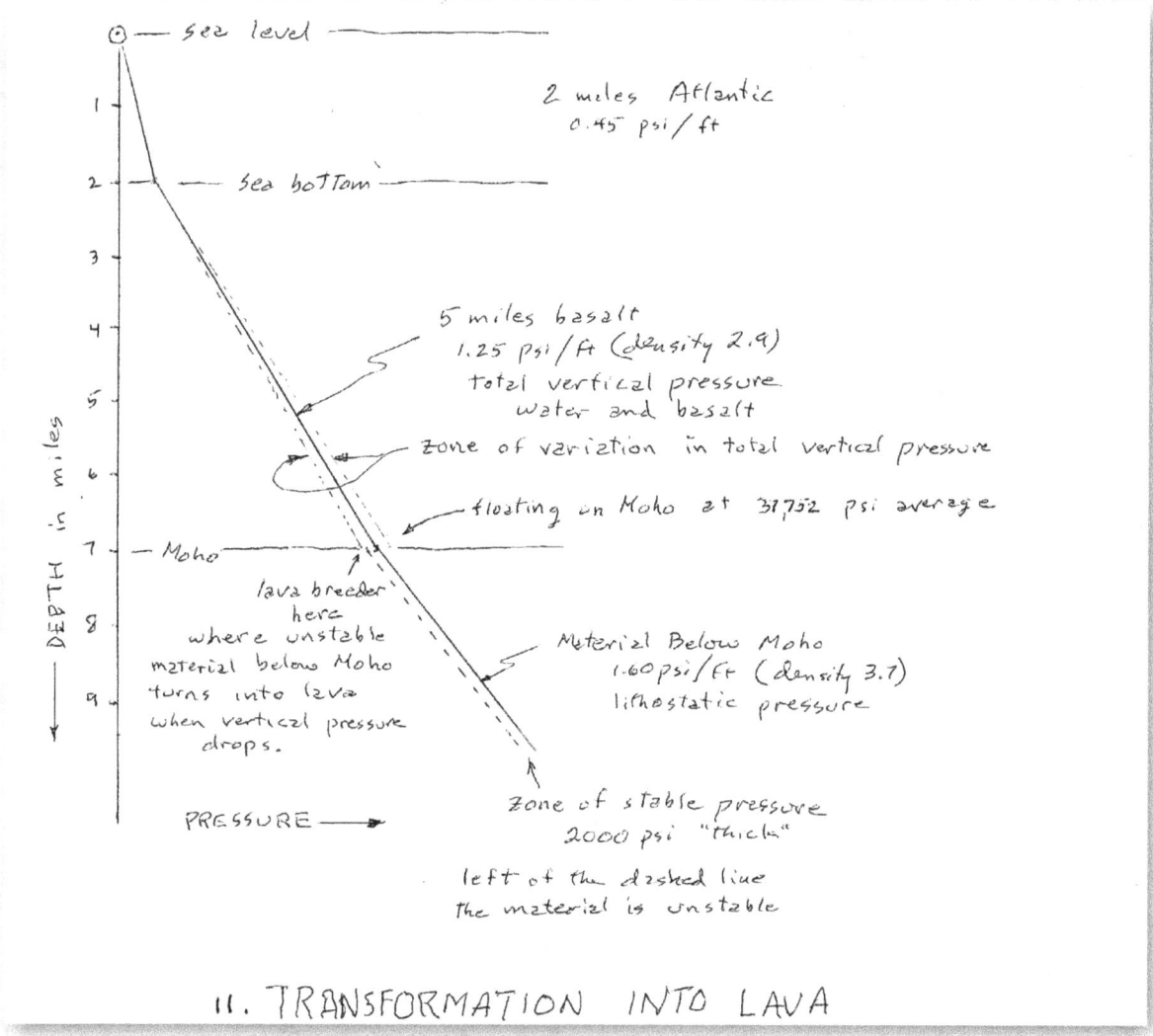

Figure 11. In the mid-Atlantic spreading zone the buttressing of the north-south prisms results in local variations in total vertical pressure. Using the assumed pressure gradients, the floating pressure on the Moho is calculated to be **37,752** psi. If the threshold of stability were negative 2000 psi, the diagram shows the range of stability pressures. Lava would be released where the pressure on top of the Moho drops below **35,752** psi.

The estimated threshold for lava formation, of negative 2000 psi is quite arbitrary. The threshold could just as well be 1000 psi, 100 psi or less. Even 100 psi pressure is very substantial, equal to the pressure of more than 200 feet of water.

P. LAVA BREEDERS

The locations just below the Moho where the pressure on top of the Moho is less than the stability pressure below the Moho, "lava breeders" form, and lava is released. These locations shift with time. Stress is re-distributed when the released lava intrudes into zones of weakness in the basalt, pumping up and healing the fractured basalt. The effects of the "deeper buttressing" and "shallow leaning" are additive, resulting in shifting locations of "lava breeders", which are oriented north-south, and occur just below the Moho.

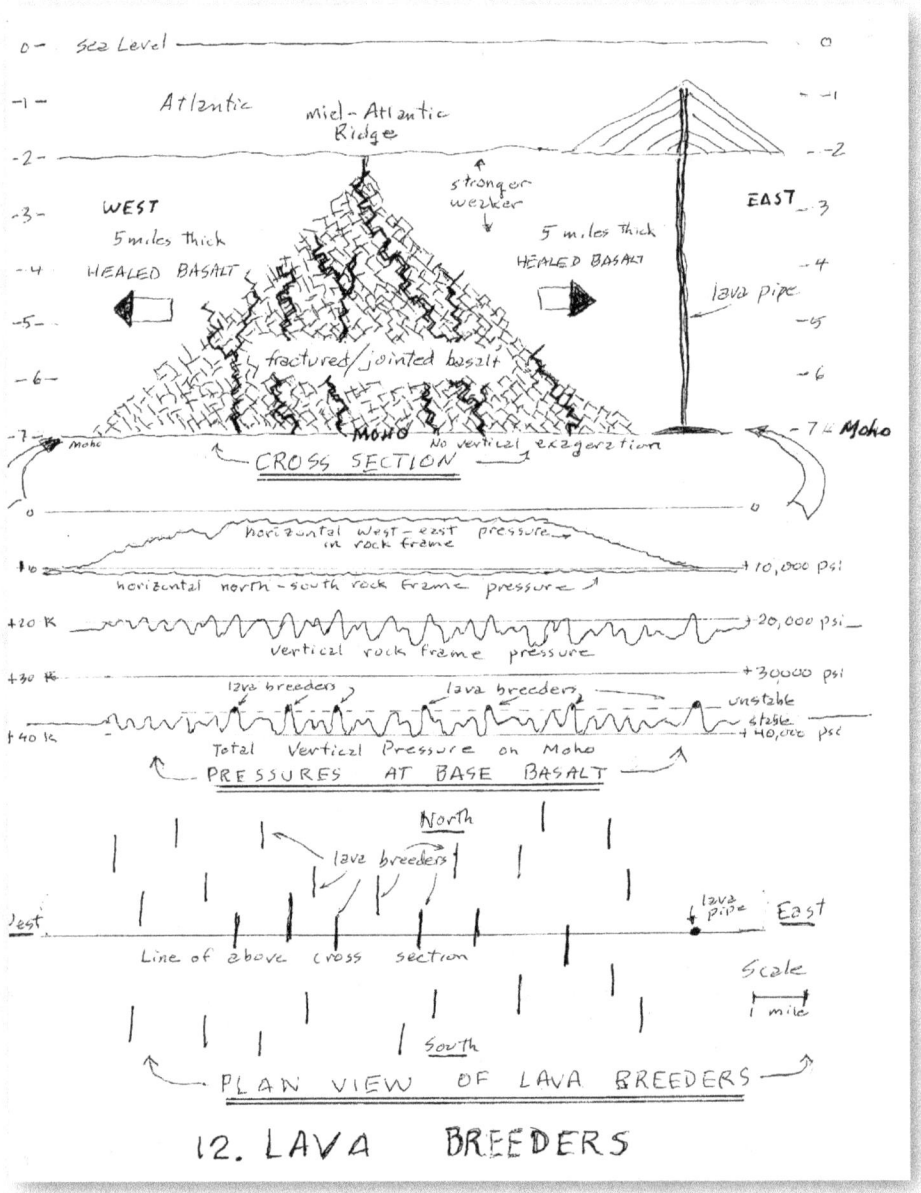

Figure 12. This diagram is actually three interconnected diagrams in one.

The **top** cross section shows the fault/joint system being injected by lava from six lava breeders. Also one lava pipe is shown feeding a seamount.

The **center** display shows the pressures/stresses at the base of the basalt, i.e. on the Moho. The vertical scale is from 0 to 40,000 psi. The lowest pressure profile shows the total vertical pressure on the Moho. This pressure is centered at the calculated floating pressure of 37,752 psi. Pressures less than and greater than 37,752 psi balance. Assuming lava will form if the pressure differential in the material below the Moho is 2000 psi below the floating pressure, a dashed line at 35,752 psi is drawn. Above this line at pressure less than 35,752 the material below the Moho will be unstable, and will transform into basalt lava. Note that the pressures at the six lava breeders (plotted directly below the breeders in the top diagram) are low, in the unstable range of the material below the Moho. Also graphed is the vertical rock frame stress, from which the horizontal rock frame stresses are estimated, based on a Poisson's ratio of 0.25. To compute the total horizontal pressures, add the water pressure to the directional rock frame stress.

The **bottom** display is a plan view of the Moho, showing the line of the top cross section and the individual lava breeders just below the Moho.

Q. CHEMICALLY EQUIVALENT?

Next question: What is the evidence that oceanic basalt and the material below the Moho have the same chemical composition?? Let's go half way around the world to the Pacific Rim. Around the Pacific, oceanic crust is being subducted under the surrounding continents, in many locations. The slabs of subducting oceanic crust are pushing their way down into the mushy material below the Moho. This push is fueled by gravity, which causes the plate to slide off the high Moho under a mid-oceanic ridge. A brittle slab of basalt, several miles thick, follows down the channel cooled by prior passage of the slab. The Benioff curtain of earthquake hypocenters tracks the descending slab.

The speed of the descending slabs is known from the width and age of the magnetic stripes on the ocean bottom. There is an established scale of the ages of magnetic reversals for the submarine magnetic stripes found under oceans. Speed of descent is measured in inches per year.

The subducting Pacific plate spends millions of years in the cooled channel before reaching a depth where earthquakes cease. "No earthquakes" signifies that all the brittle basalt has been transformed into the mushy material below the Moho. This transformation shows that the material below the Moho has the same chemical content as oceanic basalt.

How is it possible for one material to have different densities? You might ask that of the water-ice system.

R. HEAT OF LIQUEFACTION

Below the Pacific, subducting oceanic basalt spends millions of years transforming to the mushy material below the Moho. Why did this take so long?? Under the Atlantic apparently the material below the Moho gave off a substantial **"heat of liquefaction"** when lava was formed. Under the Pacific the opposite reaction takes place, and it takes a long time for subducting basalt to acquire that much energy.

Under the Atlantic there was a change of state from more-dense material below the Moho to less-dense basalt. Reversing that change of state from basalt to the material below the Moho required that energy had to be added under the Pacific. Energy in the **heat of liquefaction** stems from three sources. First there is the pressure (potential energy). Second, there is the heat energy, represented by the temperature. Third, "other energy" possibly from direct electric current (telluric currents) occurring naturally, contributed to the stored energy in the "soup". The difference in density across the Moho dictates that the molecular/ionic material is compacted below the Moho. Replacing the **"heat of liquefaction"** apparently took millions of years as the crust descended hundreds of miles at speeds of inches per year.

When the **heat of liquefaction** is released under the Atlantic, the temperature of the lava greatly exceeds the temperature of the surrounding material above and below the Moho. But as the hot basalt solidifies above the Moho, it does not change state back to the material below the Moho, despite having high temperature. Therefore it is assumed that more than heat energy is required to cause the transformation from lava/basalt to the material below the Moho. "Other energy" (naturally occurring direct electrical current?) is the missing additional source of energy.

S. RING OF FIRE

We know from the configuration of the Benioff zones, that subducting oceanic crust descends down at an angle from 20 to 80 degrees from horizontal. The fracture system in the miles-thick slab of subducting brittle basalt must be saturated with sea water. Movement of the plate of oceanic crust, generated by gravity-sliding off a mid-oceanic ridge in the Pacific, carries the descending oceanic crust down the cooled channel within the mushy material below the Moho. When the water-logged plate reaches a depth of several tens of miles, volcanic eruptions, rooted alongside the descending plate, create the tall onshore volcanic peaks of the Pacific "Ring of Fire".

Assumption. There are a few hundred volcanoes in the Pacific Ring of Fire. The annual amount of lava released from all of these volcanoes put together is small compared to the volume of annual subduction of all the plates of oceanic basalt sinking under the edges of the shrinking Pacific.

What causes these volcanoes?

"Lava breeders" similar to those under the mid-Atlantic ridge, may also be the source of lava for the volcanos in the Pacific Ring of Fire " Lava breeders" occur where the pressure down on the Moho is less than the stability pressure of the material below the Moho. What could cause such local deficit pressure on the Moho around the Ring of Fire?

One explanation has to do with the relaxation of stress inside the miles-thick subducting slab of oceanic crust as it clears the interference from the adjoining slab of continental crust. Here the slab straightens out, and proceeds deeper into the passive material below the Moho. See Figure 2 where the subducting slab bends for the second time after passing beyond the jamming effect of the adjacent continental crust.

If the Moho is defined as the change in velocity from crust to the material below the Moho, then in the case of a subducting slab in the Pacific, having cleared the obstructing continental crust, there are three Moho interfaces: the base of the continental crust, and the top and bottom of the subducting slab of oceanic crust

As the descent of the subducting slab of oceanic basalt straightens out (the second bend in figure 2), the axis of this bending is horizontal. This axis of bending is parallel to the intermediate stress in the rock pressing down on the Moho. The maximum pressure down on the Moho is the vertical load of all the rock, water and atmosphere above. The minimum principal stress is perpendicular to the other two, roughly perpendicular to the subduction surface.

As the miles-thick subducting slab of oceanic crust is straightening out, thereby tending to equalize the internal bending stress, the bottom layer of the subducting slab will experience relaxation. Differential movement across fractures and joints at the base of the relaxing part of the slab will result in local differences in vertical loading on this Moho. This is the same condition assumed to enable lava breeders in the mid-Atlantic ridge. (See figure 11)

In a subducting slab under the Pacific Rim where a local difference in loading on the Moho exceeds the stability condition there (see figure 8), a "lava breeder" will form, generating a horizontal subterranean lava pool just above the Moho. As the pool increases in diameter, eventually it will take less energy for the lava's buoyancy pressure to break through the overlying materials, forming a vertical pipe leading up to a volcano in the Ring of Fire.

If this hypothesis is correct, each volcano of the Pacific Ring of Fire should be located above the second bend (figure 2) of a subducting slab of oceanic crust.

T. SUBDUCTING PACIFIC BASALT TRANSFORMED

After the volcanism of the Ring of Fire occurs, the slabs of oceanic crust continue descending further down into the hot material below the Moho. As a slab descends, the outer layers of the miles-thick slab peel away, outer layer by outer layer, as the **heat of liquefaction** is added to the exterior of the subducting oceanic crust. The outer layer of the basalt slab changes into, and is mixed with, the surrounding mushy material below the Moho. The transformation progresses inwards into the slab. Below a depth of about 450 miles the earthquakes cease, signifying that the entire remaining brittle inner core of subducting basalt slab has been transformed into mush. Moving at a rate of inches per year, the basalt spends millions of years in the pre-cooled channel before completing the transformation. The distance down along the path of the brittle descending basalt is as much as hundreds of miles as measured by the distribution of earthquake hypocenters in the Benioff zones.

U. SEEKING ANALOGY

How can we believe **Moho Motion's** theory of changed states, bridged by the **heat of liquefaction?** A good analogy would be help sell the concept. But there isn't one known to your author.

In the case of water, we know heat must be added at constant temperature while water transitions from ice to water, and again at constant temperature from water to steam. Our human experience feels comfortable with a system of ice-water-steam, but actually the H_2O phase diagram is far more complex with a dozen components. Google it! Moral: nature will surprise you, and you are naïve if you expect that your experience and intuition trump reality.

During the time while a small chunk of basalt acquires the heat, pressure and "other energy" needed to transform it to the material below the Moho, its temperature will remain essentially constant, or at least change temperature more slowly. When enough heat, pressure and "other energy" have been assimilated, the small chunk will complete the transition to the stability condition of the material below the Moho. It's a little like an ice cube transforming to a puddle of water.

We are used to the three states of water, which is a single chemical (H_2O). The material below the Moho is not a single substance, but rather a mat of dark mineral crystals, with different chemical compositions, in a high-energy "soup" of ions. There is no single chemical compound, yet the material changes state. Your author knows of no analogy supporting the validity of such chemical behavior. But the geological evidence is there. The Benioff zones around the shrinking Pacific all fizzle out at depth, indicating that the slab is transformed, and is no longer brittle. Transformation of the oceanic slab of basalt apparently did occur.

V. ATLANTIC BASALT TOO COOL TO TRANSFORM

The basalt of the North American plate under the Atlantic also received heating from below, as it slid over the Moho. Does basalt transform under the Atlantic? It has been floating on material with equivalent chemical composition since the Atlantic parted 150+ million years ago. Has its bottom surface started to transition, like the subducting basalt under the Pacific?

The temperature on **top** of the oldest Atlantic basalt has been cooled by nearly-freezing sea water during its long trek from the spreading center. This cooling has been going on ever since the pre-Atlantic continent parted, and the resulting Atlantic depression filled with sea water. Radiant heat arriving at the base of the basalt probably bled off upwards in response to the cold temperature above. Therefore, it is unlikely that transformation of basalt occurred under the Atlantic.

A further line of evidence reinforces this conclusion. Subducting slabs of oceanic crust around the Pacific are enclosed **in** the material below the Moho. Under the Atlantic oceanic basalt rests **on** the Moho. This suggests that necessary conditions for the transitioning of basalt are 1) being surrounded by material below the Moho, and 2) movement deeper down into the material below the Moho. These conditions exist only below the Pacific.

W. EVIDENCE THAT MATERIAL SUB-MOHO HAS NO STRENGTH

Around the Pacific, subducting plates of oceanic basalt become fractured when experiencing deformation from two periods of flexing: 1) deflection down after colliding with the thicker adjoining continental plate (or an adjoining oceanic plate), and 2) straightening out in the mushy material below the Moho once the obstruction is cleared (see Figure 2).

This straightening is evidence that the material below the Moho has no strength. Bending stresses in the deflected slab of rigid basalt equalize, causing the slab to straighten when the surrounding material below the Moho offers little or no resistance to the straightening.

The descending rigid slab is less dense than the surrounding material below the Moho, but like a sheet of buoyant plywood released edgewise into a swimming pool, it sinks. Besides, the slab is being pushed by the rest of the plate sliding off high Moho.

X. SUBDUCTION

When subduction first starts around the Pacific, the brittle slab of deflected basalt invades and straightens out into the material below the Moho. Room for the massive intrusion of oceanic basalt would be created by hydraulic displacement of, and massive recrystallization in, the mat of crystals below the Moho. After the initial introduction of the slab of oceanic basalt below the Moho, further movement down the channel would simply "follow the leader".

Once and a while subducting oceanic crust finds that less energy is required to start afresh, by forming a new channel of descent. For example, subducting oceanic crust may grind off so much of the confronting continental crust that the deflection point moves further inland. Then it may require less energy for the oceanic crust to leapfrog the old channel and start a new channel. That may leave the oceanic slab in the old channel, cut off and hanging, without a sliding plate to push it along

In California there is evidence for successive ages of volcanos in the Rim of Fire. Oldest is Morro rock on the central coast, next are the Buttes near Sacramento and currently there are the three active volcanos: Shasta, Lassen and Mammoth in the Sierras. Of course, the confusing deformation along the San Andreas Fault/Sea of Cortez alignment complicates this simplification.

Y. ANATOMY OF A SPREADING CENTER

Under the mid-Atlantic ridge there are hundreds of individual north-south spreading centers. Some spreading centers are less than a mile long, and others exceed tens of miles in length. Transform faults offset the separate spreading centers (at right angles).

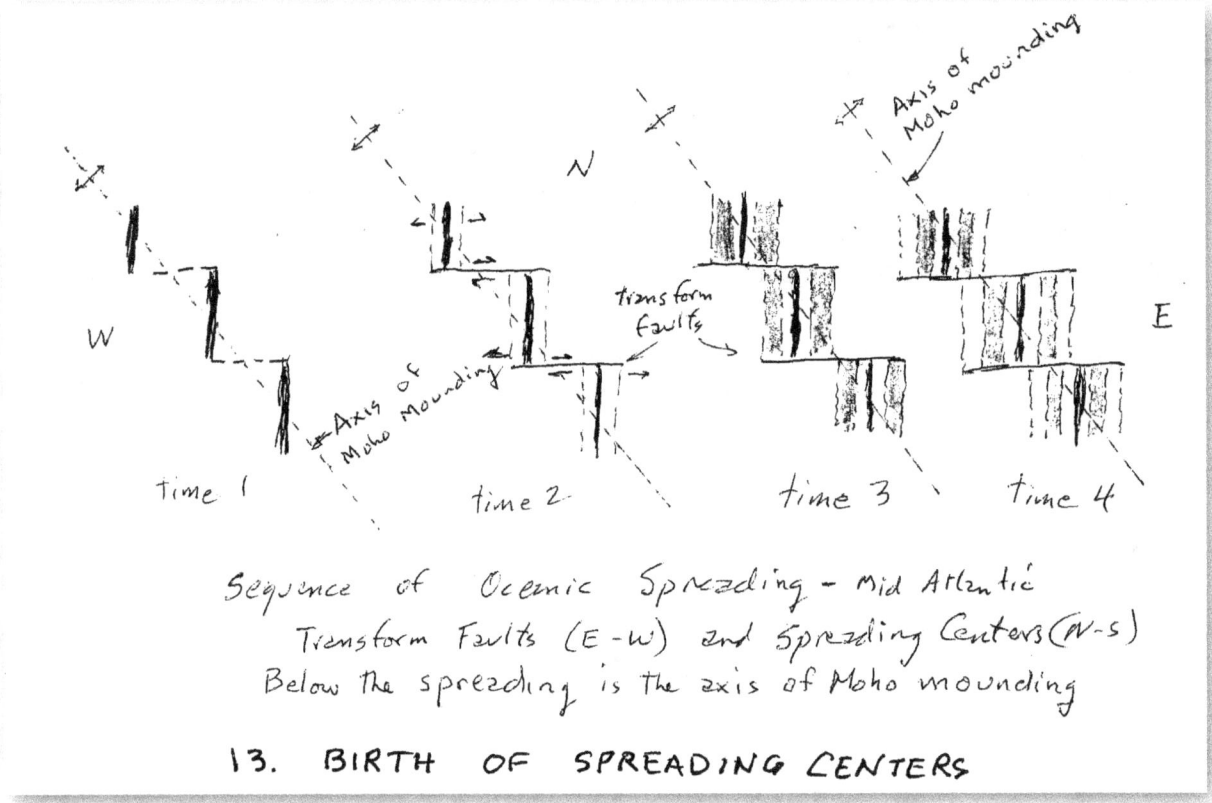

Figure 13. Transform faults offset spreading centers. Below the spreading centers is the axis of the rising ridge in the Moho. The mounding of the Moho initiated the spreading of the Atlantic. The two overlying plates are constrained by their neighboring plates to move only west or east.

How was the structure of the mid-Atlantic ridge initiated? One explanation goes like this. The ancient continent originally was unbroken when the Moho started rising as a meandering mounding. This meandering rise was destined to become the mid-Atlantic ridge. The ancient continent broke into two plates. The two new plates started sliding, west and east, off the high Moho ridge. But the crust could only move in the direction of least resistance, namely where the presence of its neighbors allowed movement. Lazy Nature accommodated this economically with transform faults offsetting spreading centers. The vertical transform faults, extending down to the Moho, are miles in height. The cold brittle uppermost part of the oceanic crust, just under the sea bottom, acts like a horizontal splint to keep a plate tied together. Once started, these transform faults continue to accommodate movement through time, as indicated in figure 13. Note that the location of, and movement on each transform fault was set when the pre-Atlantic continent broke up, coinciding with emplacement of the first mid-oceanic basalt

The undefined mechanism operating in the material below the Moho created the meandering elevated Moho. The mechanism also pushed up enough additional material under the Moho to feed the creation of all the new oceanic crust in the mid-Atlantic. "Slow motion" seems appropriate to describe the mechanism controlling the mounding Moho. It has been in operation without much change for a couple hundred million years.

The two plates slid away independently, if need be at different speeds, depending on the constraints to the motion of each plate.

You might observe that the shape of these transform faults in Figure 13 is like a series of parallel knife cuts across a French baguette. Bon appétit!

Z. LITHOSTATIC PRESSURE IN FLUIDS

Lithostatic pressure in the "soup" within the material below the Moho, is similar to lithostatic pressure in sedimentary rock. In flat lying sedimentary rocks, lithostatic fluid pressure invariably is associated with membrane activity. Permeability low enough to preserve such high pressure does not exist in rocks. Professionals: pencil it out! Analogy may be useful here. If lithostatic pressure in sedimentary rocks is sustained by membrane activity, let's consider a similar cause for lithostatic pressure in the material below the Moho.

AA. PLATE MOTION

Motion of the North American plate is influenced by: a) areas of higher or lower head below the Moho, resulting in mounds or swales in the shape of the Moho, over which the plate moves, b) resistance to internal deformation of the plate, especially in the shallower, stronger part of the crust, c) loading on the plate (sediments, glacial, water etc.), d) lateral plate boundary constraints, and e) "crashing" into the Pacific plate.

A plate moving over a raised bump on the Moho will experience lateral relaxation as the plate attempts to minimize elevation gain. This lateral relaxing will cause horizontal bending with the axis of bending in the direction of plate movement. This bending axis is, by definition, the same direction as the intermediate stress. Lateral relaxation, as the plate moves across the crest of mounded Moho, may lead to formation of horizontal lava breeders and the eruption of flood basalt.

A plate moving over a local depression in the Moho will sink into the swale

BB. ICELAND

Art Meyerhoff insisted that old fossils found in sediments on the surface of the island did not fit the new paradigm of plate tectonics. The island lies directly in the middle of the mid-Atlantic ridge. So, according to the new plate tectonics theory, the island should be split in two, with youngest basalt in the middle.

Art was right. Ten million year old sediments did not belong above a spreading center. But in the enthusiasm with which the new paradigm was greeted, Art's objection was swept under the rug.

In **Moho Motion** the theory is that lava breeders occur just below the oceanic Moho where the overlying oceanic crust is being pulled apart. Under continental Iceland the Moho is considerably deeper than under the adjacent oceanic crust. Iceland is a mini-plate of continental crust, surrounded by oceanic crust.

Going back to the origin of the Atlantic, the split in the pre-Atlantic continent did not occur all at once. Rather, the split progressed to the north. When the splitting got to Iceland, for some reason the lava breeders took detours (enabled by transform faults) around the mini plate. Offset by transform faults, one half of the spreading ended up on the west side of the island, and the other half on the east side. The two departing plates were held together by the strength of the top layers of oceanic basalt. As new basalt erupted from lava breeders just below the base of oceanic crust in spreading centers located east and west of the island, the basalt attached to, and became part of, each departing plate. Some of the basalt leaked up into the upper part of the continental mini plate, and from there up onto the island's surface. This condition continues to the present.

Around the edges of Iceland the shallow oceanic basalt is not pure, as in the rest of the Atlantic spreading centers. Erosion on the island contributed sediment, which became interbedded with shallow basalt there.

Fourteen

A Geologist's Prejudice

A little digression here may offer a welcome break, and may shed some light on geologists' mind sets. We will get back to business in the next section, but for now it might be useful for the laymen still with us to peer into a geologist's mind to gain perspective. Young geologists are introduced into an inexact science where common sense and observational power are meshed. They also inherit a culture and a history of the discipline. And perhaps a little prejudice too!

Ah, to be young again! The real appreciation of geology is an outdoors thing. Glorious vistas of rugged mountains and cascading rivers are of postcard quality. The young geologist encounters these visual nuggets while learning the ropes in the profession. Not only the visual treats, but the wonderful experiences of the great outdoors -- exercise, sunshine/rain, the smell of pine needles, wildflowers dotting the landscape, the twittering of the birds, etc. beguile the neophyte. Summer boot-skiing down remnant snow patches in the high country. Bounding down scree slopes.

More importantly the geologist realizes that clues observed can be assembled to form a comprehensive understanding of the geologic material, structure and history. The understanding of geology stems from education and a good grounding in field geology. During one summer break he/she gets to know rocks first hand as they exist in outcrops in road cuts and steep terrain. What a wonderful avocation/vocation to investigate the truth of the rocks.

A few years later it may boil down to a desk and a pair of reading glasses.

A. UNINTENDED CONSEQUENCES

But the geologist's intimacy with outcropping rocks on the earth's surface has unintended consequences. The rock examined on the outcrop differs importantly from the same rock type buried in the subsurface.

The difference is what's going on in the pores in the rock. Almost all rock is porous. Pores in rock above the water table are filled with air. Below, the pores are water saturated. Chemical weathering of rock, including oxidation, biological attack, leaching, etc., occurs on the surface and down to the water table. Below the water table fluids interact with each other and the rock framework. Over geologic time this interaction can be very significant.

Geologists, conditioned to regard rock by its outcrop appearance, are slow to appreciate the effect of fluid saturation in the subsurface. The water in the pore system is an electrolyte with oodles of ions. Each ion has positive and/or negative "corners" which will attract or repel other ions.

Likewise, the interior surface of the pores is not electrically neutral. Molecules of minerals comprising the rock frame also have charged "corners" which interact with fluids in the pores.

B. MEMBRANES

Voila! We have the makings of a membrane! Membranes are selectively permeable to water and ions. An entire industry for desalinization of water is built on such selective permeability. The forms of energy gradients across membranes in the subsurface are many, including pressure, electric potential, temperature, biological activity, fossil energy in trapped fuels, chemical reactions and ion concentrations. Membranes in a static system can balance contrary forms of energy, creating a local standoff.

In an early experiment proving the concept of desalination, a fine grained shale rock was ground up and reconstituted as a strong wafer. A pressure difference of 5000 psi on salt water on one side, produced a trickle of fresh water coming through the small wafer. The wafer did not allow salt ions to pass through.

Fine grained rock can behave like a membrane. So what? Why should geologists pay attention to membrane behavior, when the geologist has seen, hefted, hammered, scratched, selfied and developed a familiarity with the particular rock in its exposure at ground level? The reason is that membrane behavior can change the character and performance of the combined totality of rock and fluid buried in the subsurface.

C. IRREDUCIBLE WATER SATURATION

Membrane insights should be well received by geologists in the petroleum business. Petroleum engineers and well-site geologists routinely take side wall samples and core samples from petroleum zones in wells. Over a million of these core samples have been tested for irreducible water saturation, among other things. The point of this test is to find out how much oil can be crammed buoyantly into the pores in the rock at any given height above the petroleum/water contact. As buoyancy pressure increases, the film of water on the walls is squeezed thinner. When the film of water coating the sides of the pores cannot be squeezed any thinner by increasing pressure in the oil, the maximum oil saturation has been determined. Petroleum engineers need these data to compute reserves of petroleum. At this point the thickness of the water film on the walls of the pores is controlled by molecular forces rather than hydraulic (buoyancy) pressure.

D. SPONTANEOUS POTENTIAL

Another membrane-related phenomenon, well known in the oil patch, is "spontaneous potential". After an oil well is drilled, a sonde is lowered down to continuously measure the electric potential of the layers of sediment intersected by the well. The range of potential is a few millivolts. The electrical potential of sand

and of shale is consistently and clearly different. The electric log identifies lithology and aids in correlating the stratigraphic sequence encountered in adjacent wells. Millions of electric logs have been run in wells drilled for petroleum and water. Spontaneous potential in water-saturated rock is membrane related.

E. PREJUDICE

But now a little prejudice kicks in. Petroleum geologists like to think of rocks as they appear in the outcrop. Membranes don't enter the picture because they are easy to ignore. Invisible and only detectable indirectly. Who pays attention? Anomalous pressures (too high and too low) are explained as fossil pressures, gas expansion, leaks from other pressure regimes, incomplete compaction, etc. Little credence is grudgingly spared for possible membrane activity.

Permeability in very "tight" shales (excellent membranes) is measured in the geophysical laboratory after first cooking off all the water and then injecting mercury, building to great pressure. Cooking off the water destroys the membrane function of the shale.

In the medical profession membranes are respected. For example, liver and kidneys clean your body fluids. Membranes at work!

F. MEMBRANE EXAMPLE

Let's acknowledge that membranes exist and function in rocks, and try to weave those insights into broader solutions for geological problems. But how can you be expected to take invisible membranes seriously without some proof?? So here goes with a ginormous example.

The gas pressure in the United States' largest gas field, Hugoton (about 75 trillion cubic feet), has always been greatly under-pressured. The original reservoir pressure of the gas was about 600 psi, whereas a column of water to that depth "should" have a pressure of 1500 psi. Gas wells were first carefully located on anticlines, following oilfield tradition. But it turned out that it didn't matter, high or low, the only conditions needed for good gas production were permeability and porosity. The head of the water in the rocks above and sideways from the Hugoton reservoir is greater than the head in the water in the pores surrounding the gas. It is bottled gas, in a huge "bottle" spanning parts of three states: Texas, Oklahoma and Kansas. Before gas production started, the walls and ceiling of the" bottle" were saturated with water at higher head than in the water- wet pores of Hugoton.

Next door in New Mexico there is another giant gas accumulation, the San Juan field, where the original reservoir pressure was much lower than it "should" have been. In the sixties, Fred Berry hypothesized that water was pumped by osmosis out the bottom of the field through shale. Osmosis reduced water pressure in the pores surrounding the trapped gas, so the gas had no way out, despite its buoyancy. Berry's explanation for the San Juan probably is related to the nearby Hugoton, too. How else could so much gas become bottled up at pressure much less than hydrostatic?

Membrane activity in rocks is under-appreciated.

❖ ❖ ❖

Fifteen

Pressure-Depth Graphs

Watching whales off the California coast, you would confidently expect that a sounding whale would be experiencing greater water pressure the deeper he dives. You can be thankful that you are not strapped to the whale, for during the dive thousands of feet deep, you would be squashed like a bug on the windshield.

And with that gripping introduction to the relationship of depth and pressure, you are about to delve into the discipline of pressure-depth graphs. As boring as they may seem, these graphs help organize our understanding of some relationships in the earth's crust. It's worth the "pain" to gain this perspective, so don't lose heart.

Pressure in the ocean depends on the depth below sea level. If the tide rises 10 feet, pressures at all depths rise 4.5 psi compared with the original condition

Hydraulic engineers simplify by referring to the **head** of the fluid system. When the fluid is motionless and its density does not vary with depth, the pressure-depth graph is a straight line. For fresh water: 4.3 psi at 10 feet, 43 psi at 100 feet and 430 psi at 1000 feet.

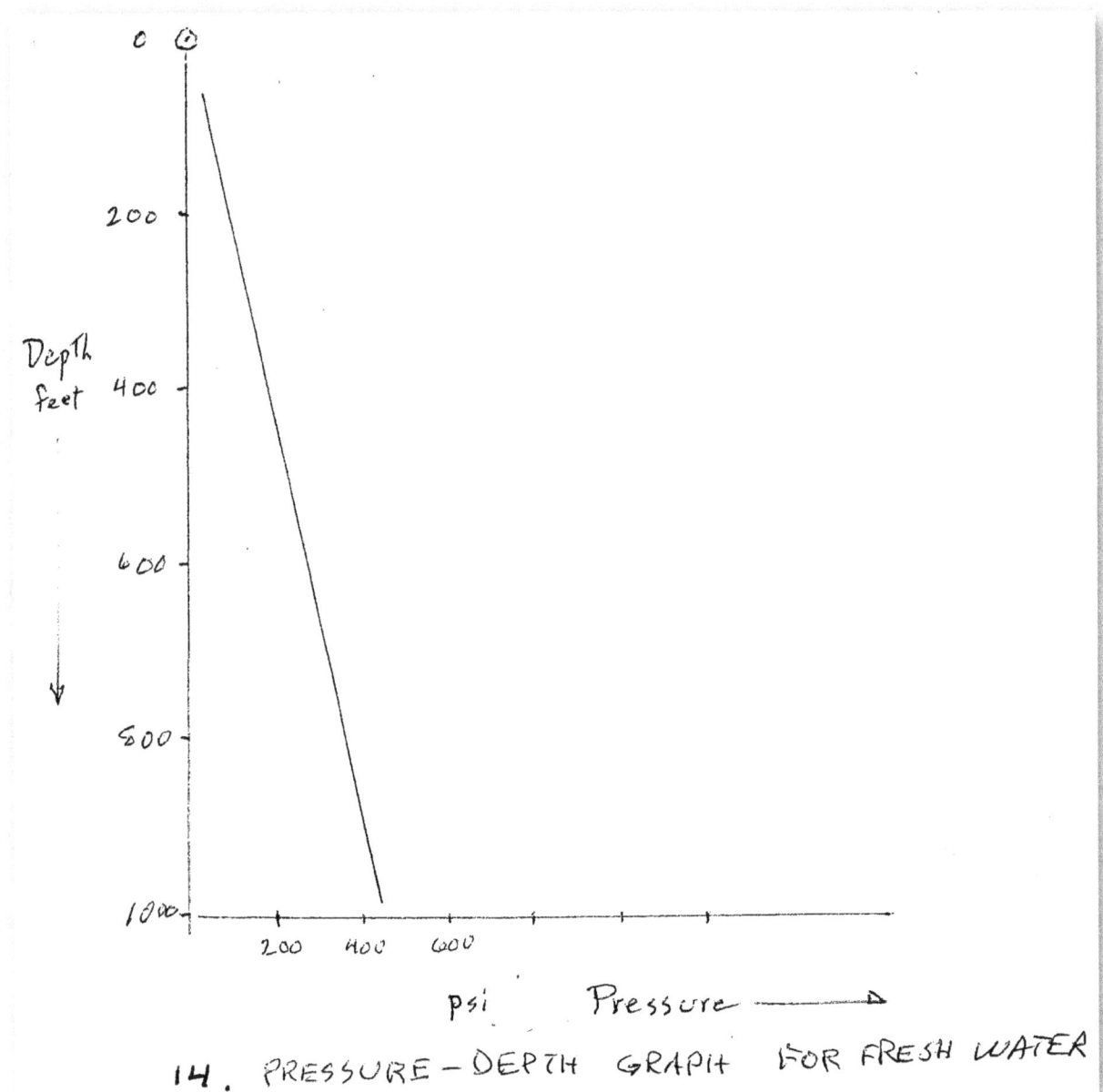

Figure 14. Pressure-depth graph of fresh water. The pressure gradient is 0.43
pounds psi per foot. At 1000 feet depth, the pressure is 430 psi.

Whether the head of a fluid is raised or lowered, the pressure-depth graph will have the same slope.
The only change will be in the elevation of the fluid-air contact, which is the **head** of the fluid. Head
trumps pressure, because head defines the potential energy of the system.

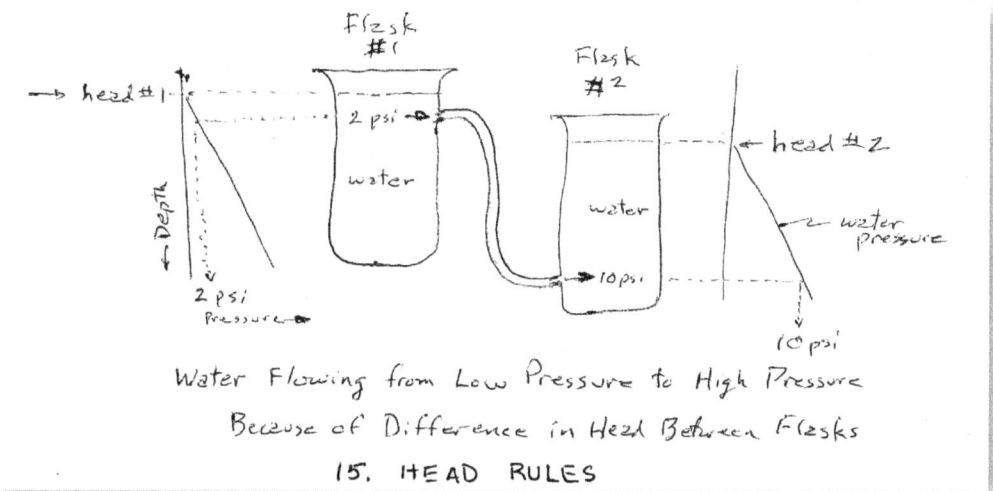

Figure 15. It doesn't seem right that fluid would flow from low pressure to high pressure. But here is a case where it does. Note the two pressure-depth graphs, one for each flask. Actually the water flows from high **head** to low. **Head** rules.

On the other hand, the real-world relationships of pressure and depth in the ocean are complex. The density of the sea water varies with temperature, salinity and pressure. Cold water sinks. Sea water moves with the earth's rotation, so the kinetic energy of a unit of "still' surface water varies with latitude. If ocean currents are not parallel to latitude, the momentum of water will change. In high latitudes tidal currents over shallow shelves can flow at speeds of several miles per hour. Shallow currents interact with deeper countervailing currents. Winds blowing over the ocean surface cause surface water currents. Changes in atmospheric pressure during cyclonic weather patterns can cause "Sandy" conditions.

But let's keep it simple, as a useful approximation. The following pressure-depth graphs assume sea water has a uniform density of 0.45 psi per foot, and fresh water 0.43. Whether in open water or in the system of interconnected pores in rock, these same gradients apply. In actuality salinity varies widely, and temperature increases the deeper you go in the earth.

Similarly, the densities of rock and lava, shown in the pressure-depth graphs here, are generalizations, based on averages of randomly selected data.

Five Pressure-Depth Categories

In a geologic context, pressure-depth relations under static conditions can be considered in five categories:

a. fluids and gases
b. massive solids, like thick sandstone
c. massive solids undergoing stress, such as in mountain building
d. over pressured and under pressured systems
e. the buoyant effect from saturating fluids on categories b. c and d

A. FLUIDS AND GASSES

Fluids and gasses, at any given location, have the same pressure in all directions: north, east, up and down. At constant head, pressure varies with depth according to the density of the fluid/gas (as long as everything stays motionless). The viscous mush, comprising the material below the Moho, fits in this category.

B. MASSIVE SOLIDS

Massive solids, like a horizontal layer of sandstone 100 feet thick, weighs down on its supporting surface. This pressure can be visualized as the weight of a column of sandstone one square inch in cross section and 100 feet tall, i.e. pounds per square inch (psi).

Note that pressure-depth curves are used to show the stress conditions of solids, as well as fluid pressures.

Simeon Denis Poisson (1781-1840) was a brilliant mathematician/physicist, who recognized that confined solid materials transfer part of an applied load sideways. The Poisson's ratio of any solid material is a primary characteristic, like density, color or hardness. The ratio varies from zero to one half (cork at 0, rubber at ½). As a practical approximation, multiply the Poisson's ratio by 2. The result is the amount of horizontal pressure resulting from vertical loading. For rubber the entire vertical pressure is transmitted horizontally. For cork, almost none. Thus the cork forced into the bottle of Cabernet Sauvignon would transfer very little of the corking force into the fragile glass neck.

For massive sandstone, Poisson's ratio may be about 0.25, meaning that gravity loading within an inch square column of sandstone would result in a horizontal stress equal to about 50% of vertical stress. This horizontal stress presses outward against the surrounding sandstone, and the adjoining sandstone presses back equally.

C. MASSIVE SOLIDS UNDERGOING STRESS

Imagine that the massive sandstone layer (previous paragraph) would be subjected to a horizontal push in a particular direction, and a geologic buttress prevented movement in that direction. The amount of that additional stress would be added in that direction to the amount of horizontal stress calculated from Poisson's ratio. Horizontal stress is now greater in the direction of the push, say north-south (both in the north and in the south directions), whereas in the west-east direction the stress remains the same as before.

Engineers refer to the "principal stresses" in a solid, oriented at right angles to each other. Stresses in a solid are defined as **least**, **intermediate** and **maximum**. In the Atlantic ridge setting, the vertical stress is the maximum principal stress, and the least principal stress is in the west-east direction of the plates pulling apart. This leaves the intermediate principal stress in the north-south direction. The direction of the intermediate principal stress is the axis of symmetry for deformation of the rocks in the pull-apart.

D. ANOMALOUS PRESSURES

Up to now it was only mentioned in passing that Mother Nature often causes the pressures in fluids below the earth's surface to be anomalously higher or lower than the "normal" pressure of a column of water to that depth. But "abnormal" pressures are common. Very common.

Most of the recent petroleum discoveries in the Gulf of Mexico were over- pressured before any petroleum was produced. In over-pressured zones the pressure in the drilling fluid must be made denser in order to safely keep water and hydrocarbons from leaking from pores in the adjoining rock into the borehole.

The upper limit for pressures in fluids in the pores of horizontally layered rocks is the pressure needed to lift all the overlying rocks and fluids. In other words, the limit for fluid pressure is lithostatic pressure. If this pressure is exceeded the fluid can lift the overburden, causing leaks. Such naturally occurring leaks tend to heal themselves as excess pressure bleeds down to hydrostatic.

In rocks being squeezed by forces of mountain building, pockets of pressure exceeding lithostatic can exist.

The diagramed pressure-depth graph in Figure 16 is from a large oilfield. Three shale sections are over pressured, and oil columns are trapped in sand reservoirs between the over pressured shales. Oil/water contacts are usually below buoyant petroleum. But here is a case where there are two oil/water contacts in each reservoir, one below and one above.

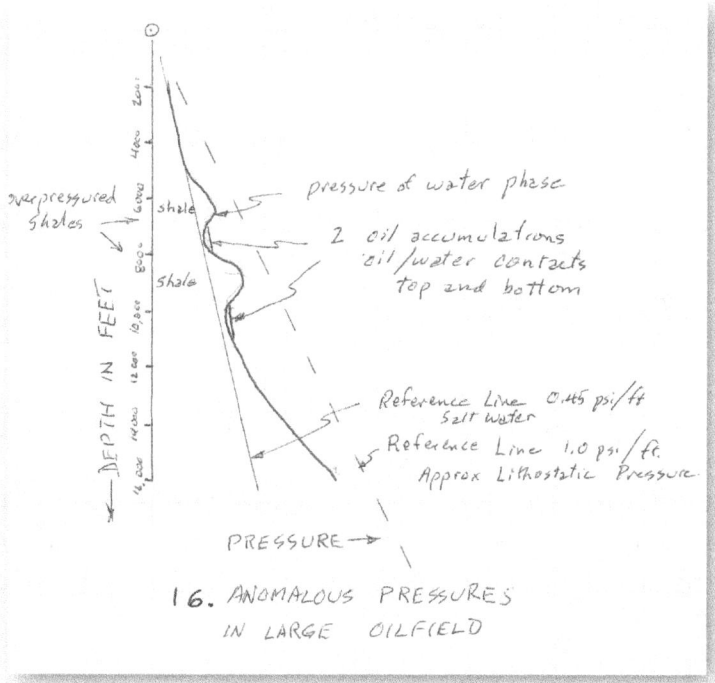

Figure 16. Two oil accumulations trapped by over pressured shales. The higher head in the shales acts as the cap. Usually shales trap because buoyant oil can't force its way into the tiny pores in the shale. But in this case oil cannot rise because of the higher head in the water in the shale. Thank you, Andy Bengtson!

E. EFFECT OF BUOYANCY

Things get interesting when there is a fluid in which the solid is immersed. Buoyancy kicks in. When you get into a swimming pool you only touch bottom in the shallow end. In deep water you hope to float. A block of sandstone weighs less in the pool than out.

Archimedes was the first to study the effect of buoyancy, more than 2200 years ago. After solving the very difficult problem of detecting diluted gold in the king's crown, in celebration of his discovery he ran through the streets shouting "Eureka", until some spoil-sport reminding him that he was running buff. At least that's the fable. On more solid ground is Archimedes Principle, in which he states that a solid immersed in water experiences buoyancy equal to the weight of the water displaced.

Karl von Terzaghi, the father of soil mechanics, was the first to appreciate (1924) that buoyancy works inside soil when the pores are water saturated. That same insight applies to rocks just as well.

The following is a little tricky, so listen up.

Fill up your ten-foot deep swimming pool with dry sand ten feet deep. The bottom of your pool supports the pressure of the sand. Now add water up to the top of the sand. Now the bottom of the pool supports the column of water plus the column of **buoyed** sand. Buoyancy reduces the vertical pressure of the sand by the pressure of the displacing water. Whether the sand is dry or water saturated, the vertical pressure on the bottom of the pool is the same.

If the sand in your pool were coarse volcanic froth ("pumice", a glass with many bubbles) with the same density as water, the dry pumice would push down on the bottom of the ten-foot pool with 4.3 psi. Adding water up to the top would allow the pumice to barely float, so the buoyed pumice would add zero psi on the bottom of the pool. Only the water pressure of 4.3 psi would press down on the bottom.

Note that the vertical pressure on the bottom of the pool is the same whether there is dry sand, or sand with water up to the top. This seems illogical because there is more mass in the pool when the pores between sand grains are filled with water instead of air. Adding water to fill the pores will add to the mass (weight) of the materials filling the pool. But our problem is to compute the pressure pushing on the bottom of the pool, not the mass of material in the pool.

Atmospheric pressure is being ignored here. Of course if you want to get picky, you have to add 14.7 psi, the pressure from all the air stacked up above us.

A few practical examples may clarify. Following are simple pressure-depth graphs. In these diagrams there is an imaginary rubber bladder which can be inflated until the pressure in the bladder just balances the pressure from the materials above. This imaginary bladder is just a mental crutch to help visualize buoyancy.

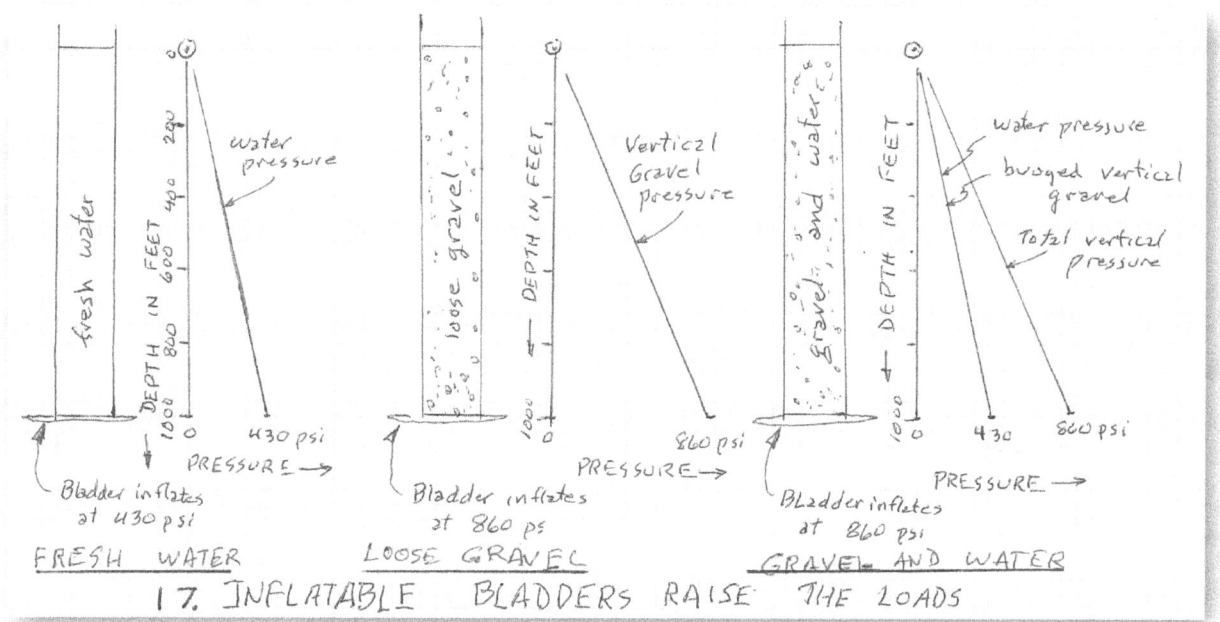

Figure 17. Inflatable Bladders Raise The Loads.

The simplest case is water. The density of fresh water is 1.0, which equates to a pressure increase of 0.43 psi per foot of depth, as shown in this figure. Pumping up the bladder just a little above the pressure of the water will make the bladder swell, pushing the overlying water up.

Next, consider the case of gravel. A 1,000 foot thick layer of loose dry gravel is diagramed. The density of this gravel is about 2.0, so the pressure gradient is about 0.86 psi per foot. Now imagine that you commence pumping up the pressure in the inflatable rubber diaphragm. Increasing the pressure in the diaphragm above 860 psi will commence lifting 1000 feet of dry gravel.

Now pour in enough water to fill the porosity in the gravel, so that the flat top of the gravel and the water level are at the same elevation. Now the pressure on the bladder is the sum of the water pressure plus the buoyed vertical pressure of the gravel. The water pressure is 430 psi for 1000 feet of fresh water, whether the water is free standing or in porosity, The buoyed vertical pressure of gravel is dry gravel pressure (860 psi) less fresh water pressure (430), equal to 430 psi. Pressure the bladder to 860 psi, and it will begin to expand. Note that 860 psi is the pressure needed to lift dry gravel or gravel filled with water to the top.

The key thought here is that in a layer of rock saturated with water from the surface down, the rock will be buoyed up by the pressure of the water. This is a simple fact, but it is hard to get comfortable with it.

❖　❖　❖

Sixteen

Drawing Pressure-
Depth Graphs

Before delving further into the wonders of pressure-depth graphs let's go through the ABCs of drawing them. First buy an 8 1/2 by 11 inch tablet of drafting paper with one inch blue fading grid divided into tenths. Next will be a small light table, say an 11 by 18 inch "Porta Trace" by Gagne, Inc. (under $100). The blue lines in the grid enable a precise lineup on a light table, so that the accuracy of the graphical projections is about equal to line width.

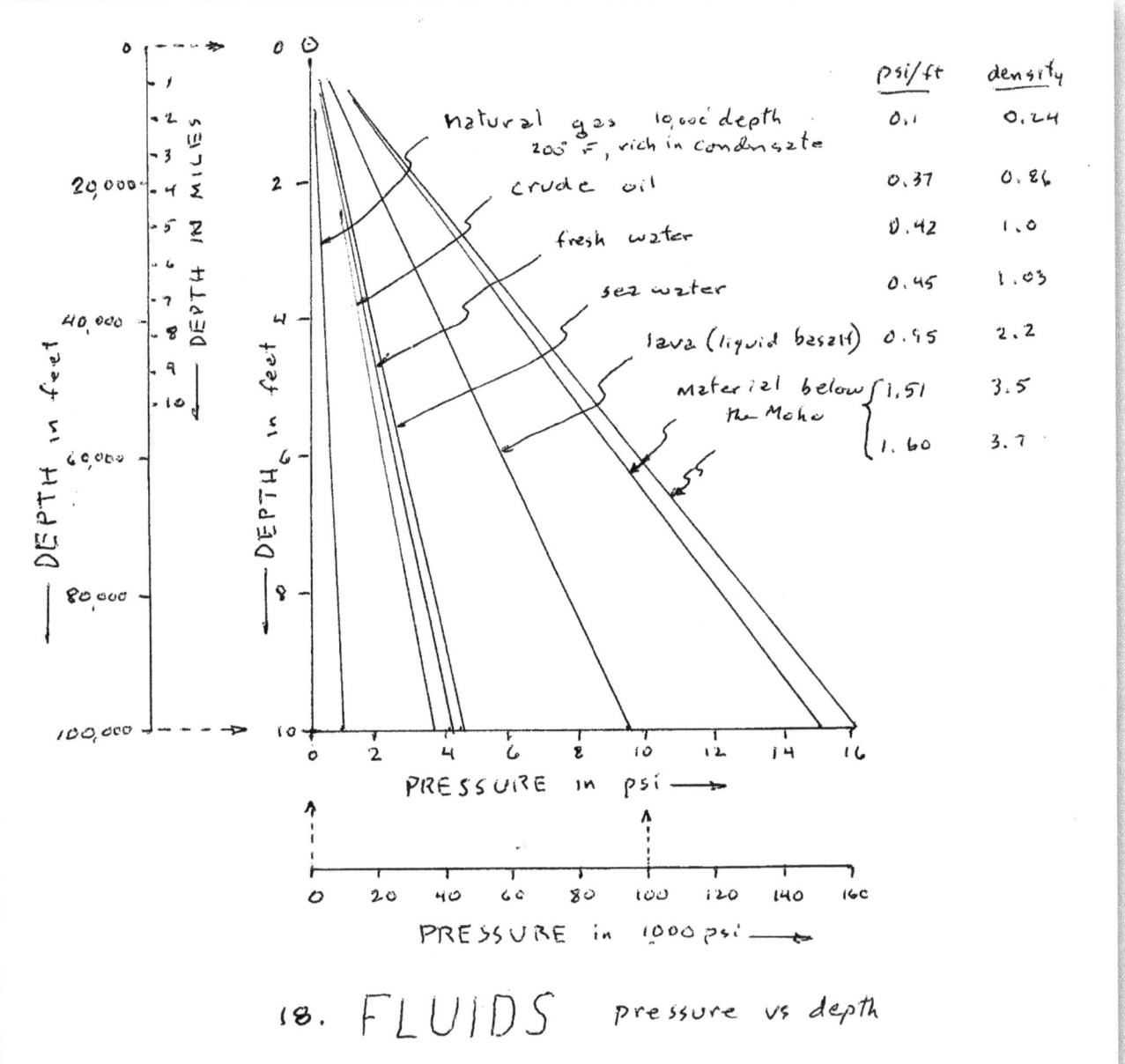

Figure 18. Fluids Master graph with pressure gradients for various **fluids.**

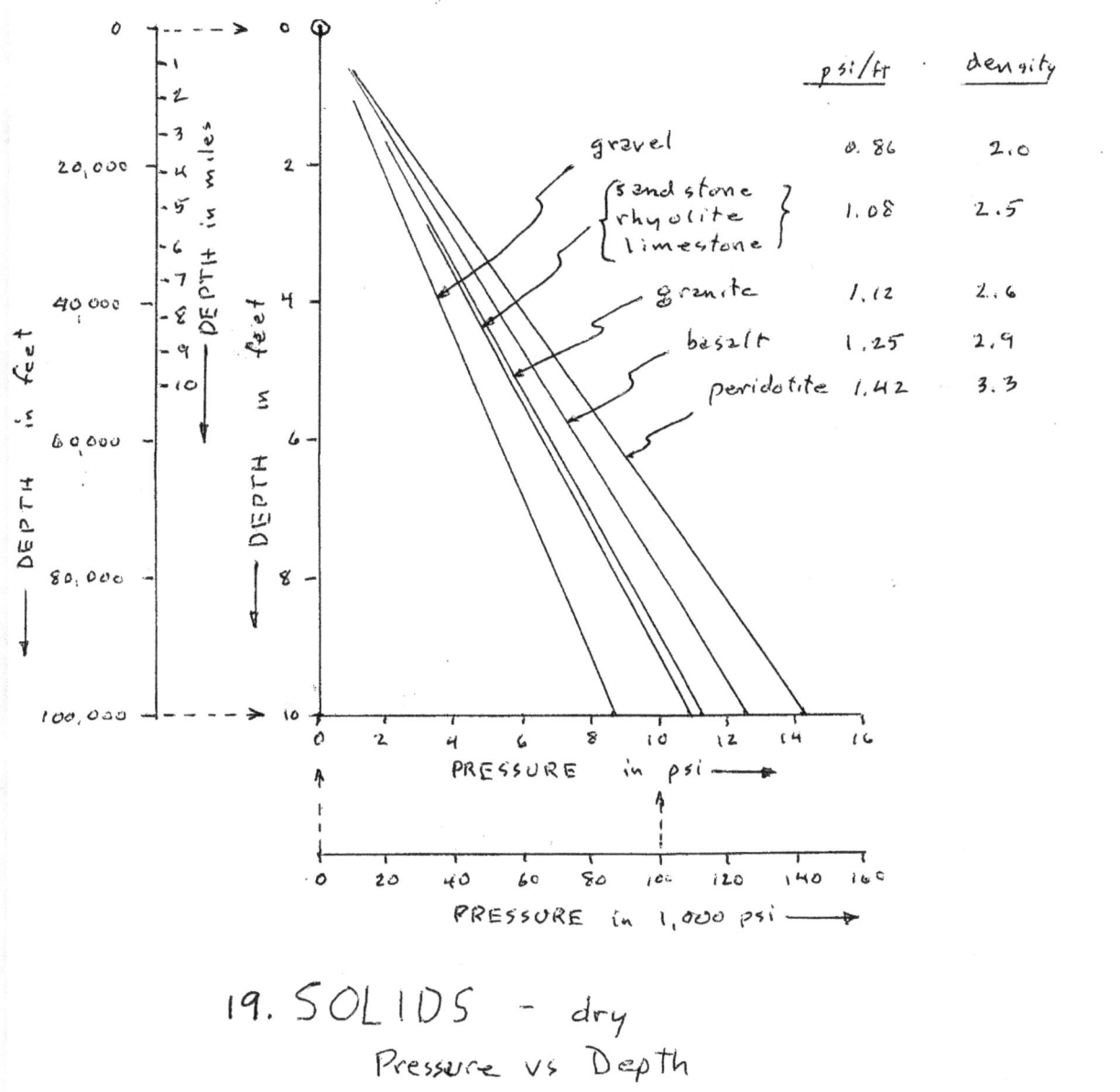

Figure 19. Dry Solids. Master graph with pressure gradients for various **solids.**

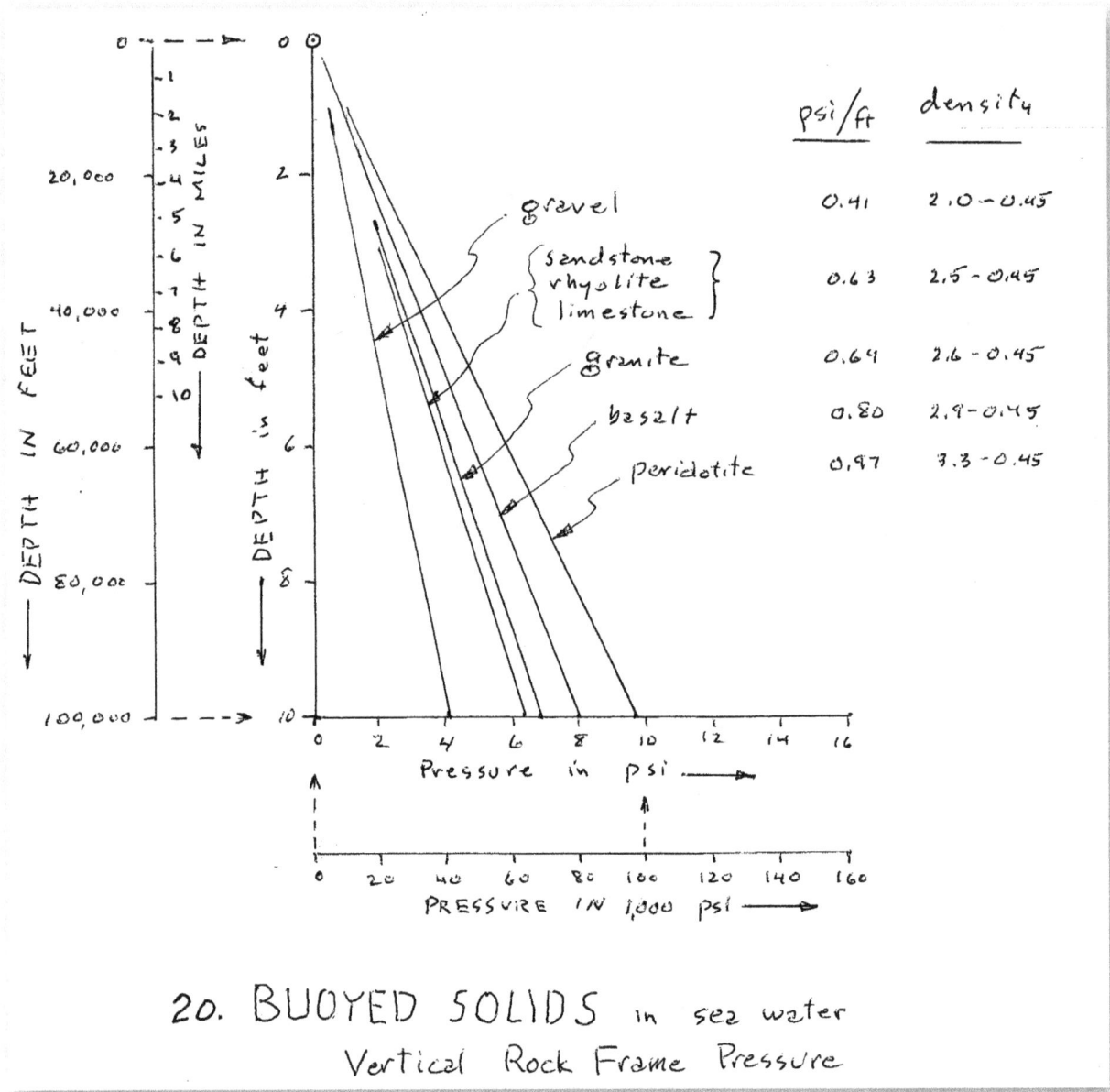

Figure 20. Solids buoyed in sea water. Master graph with gradients for various **buoyed solids**.

In all the Pressure-Depth graphs in **Moho Motion** these three graphs show the pressure gradients for three classes of materials: fluids (including gasses and the material below the Moho), solids in air, and solids buoyed by sea water. The densities used for calculating pressure gradients are averages of randomly selected published data. **Figures** 18, 19 and 20 are straight-line simplifications of density gradients. As a first approximation these are workable, even though in actuality the average densities sited are only average. At the depths cartooned, densities undoubtedly increase a bit with greater depth (unless over-pressured conditions exist). There is a lot of "play" in these data. A careful operator could improve on the accuracy of assumptions and might tailor gradients (gradually increasing density with depth, etc.).

Pressure-Depth graphs must be consistent within any one study in order for the visual impact to make sense. It is essential to maintain a constant ratio between the units of the depth and pressure. A gradient dipping 30 degrees to the right must always mean the same psi/foot gradient, no matter which depth scale is used.

All pressure depth graphs in the illustrations in Moho Motion have the same ratio between pressure and depth. Some of the diagrams show only the depth scale, leaving the pressure scale inferred. Rest assured that the pressure scale could be reconstructed from figures 18, 19 and 20.

❖   ❖   ❖

Seventeen

Lava Breeders below
Mid-Atlantic Ridge

A. VERTICAL PRESSURE ON THE MOHO

The vertical pressure pressing down on the Moho is the sum of the pressure of the sea plus the buoyed vertical pressure in the basalt.

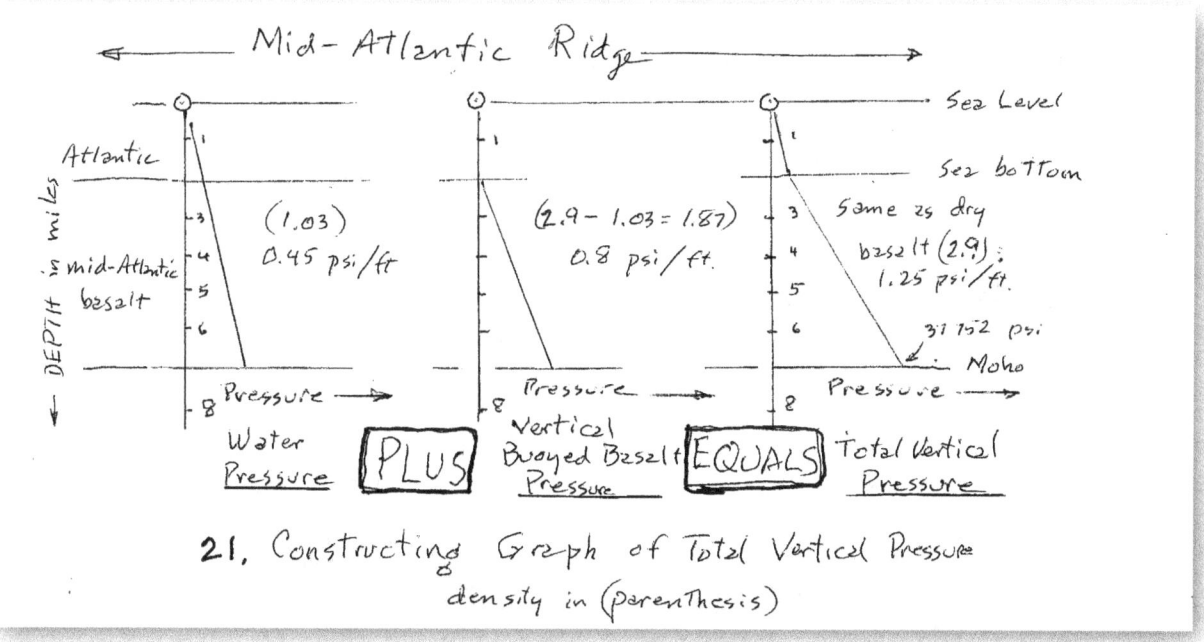

Figure 21. For the mid-Atlantic ridge area the depth of the Atlantic is assumed to be 2 miles, and the basalt five miles thick. To graphically calculate the vertical pressure on the Moho add the water pressure (0.45psi/foot) from sea level plus the buoyed basalt pressure (0.8 psi/ foot) from 2 to 7 miles depth. "The result is the total vertical pressure of 37,752 psi on the Moho." Note that the pressure gradient in the basalt is 1.25 psi for each additional foot of depth. The gradient for dry or buoyed basalt is the same. The pressure is calculated using the indicated gradients and depths.

The next two graphs compare pressure distribution under the crest of the mid-Atlantic ridge with that under the abyssal plains. Figures 22 and 23 also demonstrate how the head of the material below the Moho is graphically determined.

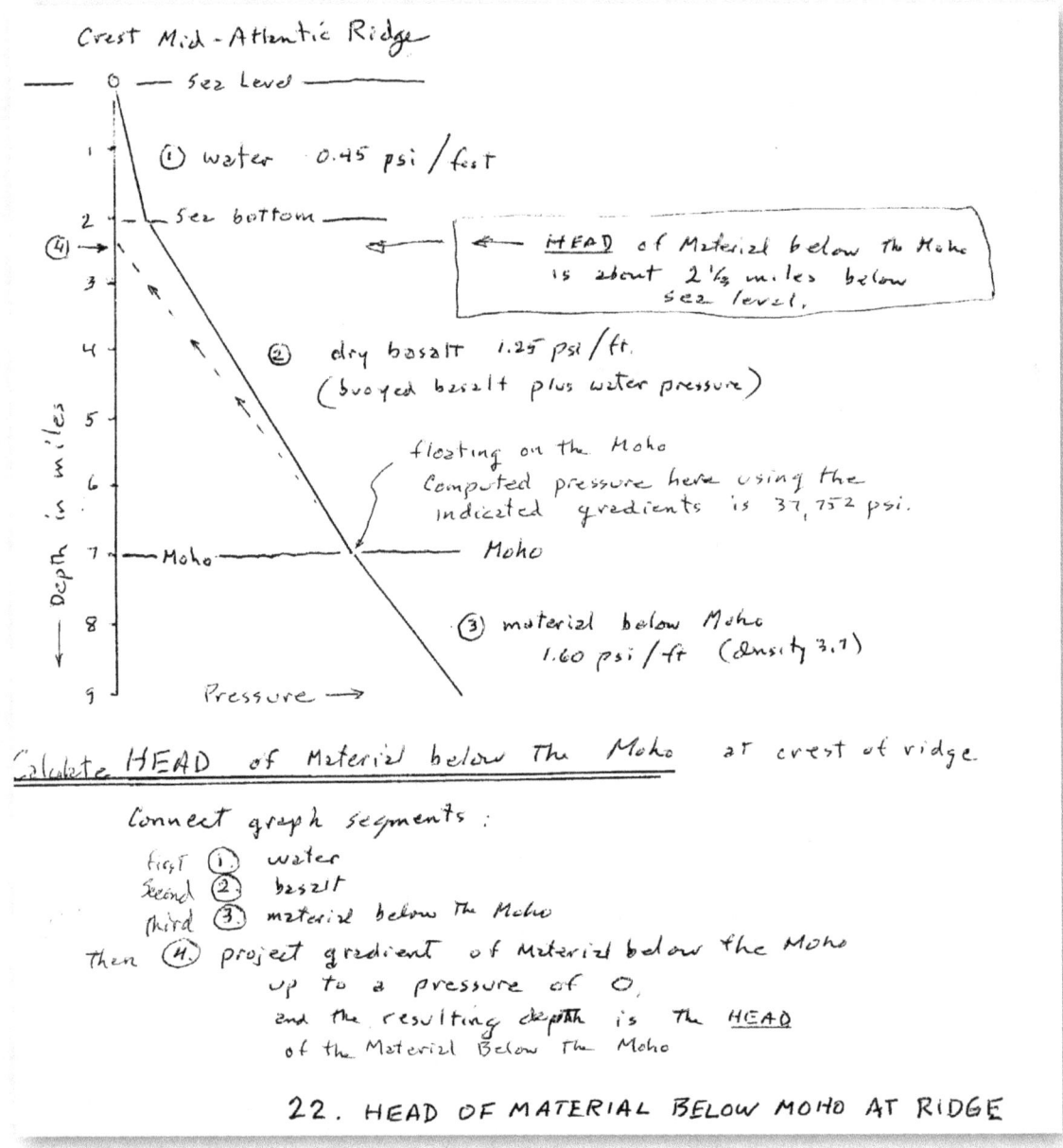

Figure 22. To chart the head of the material below the Moho in the vicinity of the mid-Atlantic ridge, start at a known pressure (sea level) and connect gradients down to the Moho depth. At that depth connect to the gradient of the material below the Moho. This is the essence of "floating". At a point on the Moho, pressure down equals pressure up. Extend that last gradient up to a pressure of zero psi. This elevation is the head of the material below the Moho (2 ⅓ miles below sea level).

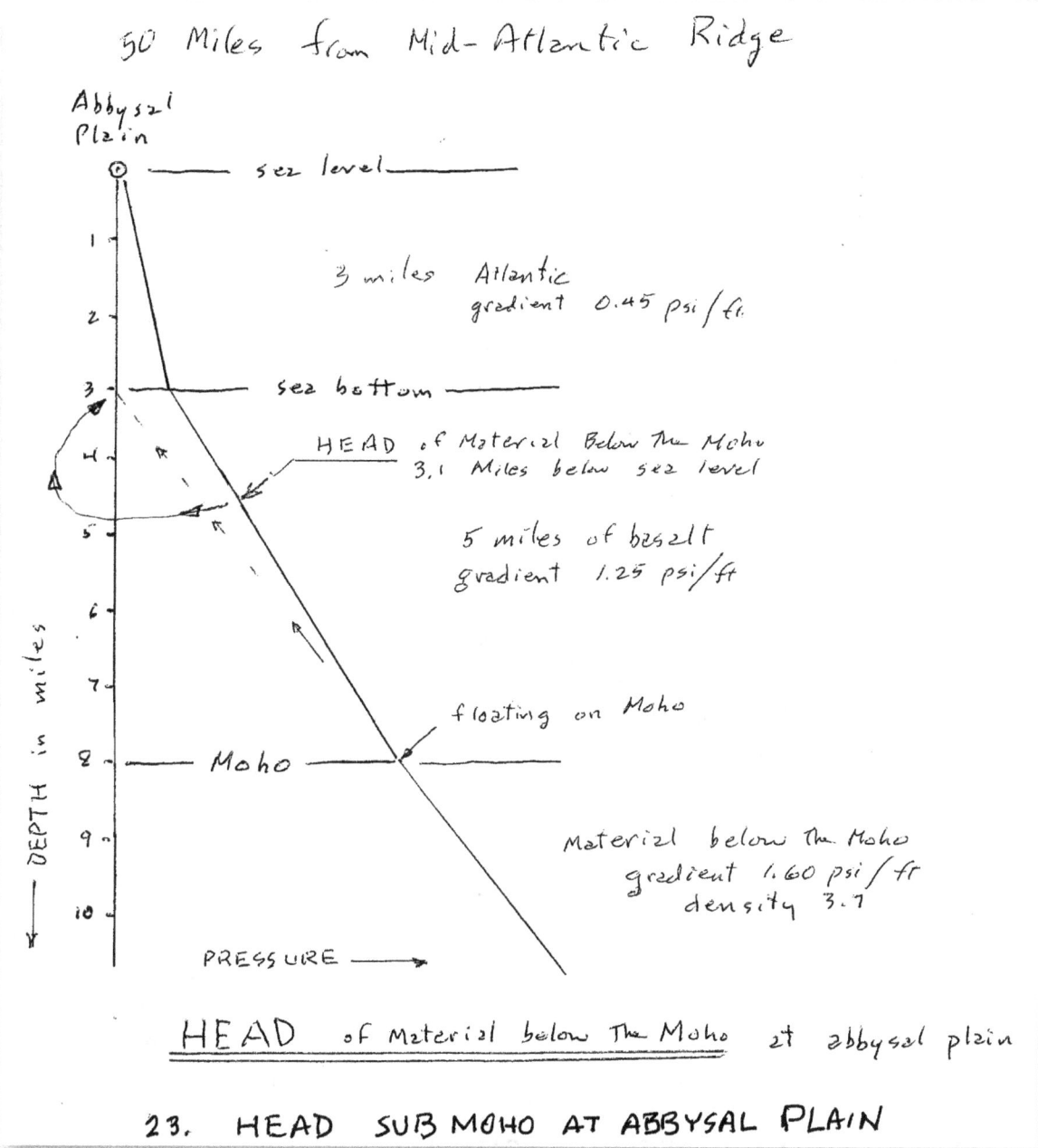

Figure 23. Using the system of figure 22, the head of the material below the Moho under the abyssal plain charts out to be 3.1 miles below sea level.

As lava is released from a lava breeder, the lava moves up through the Moho. B**elow** the Moho, in the "lava breeder" the volume of viscous material there is reduced. Reduced volume spells lowered pressure. Viscosity slows buoyant response below the Moho. Thus, formation of lava in a lava breeder below the Moho tends to prolong the lower pressure there, keeping the lava breeder active longer. It acts as a self-feeding loop.

Under the mid-Atlantic spreading centers the shape of the surface of transition, from material below the Moho to lava, probably starts as a north-south line immediately below the Moho. This line would be located below and between the edges of two fault-bounded, horizontal prisms resting on the Moho. The location of the line will trace the trajectory of least vertical pressure on top of the Moho. As the transition to lava slowly progresses, the surface of the transition will expand to a half cylinder below the Moho. As the half cylinder grows in diameter, the area of active transition will grow larger. As the area of transition increases, this will dilute access to the low pressure above the Moho, so transition will slow. Another self-feeding loop, working opposite to the first.

Combining the results of the two loops from the preceding paragraphs, it seems that initially a lava breeder will be prolific, and later it will wither.

B. TEMPERATURES OF A LAVA BREEDER

In a lava breeder below the Moho there are two temperatures. Initially there is the low temperature of the material below the Moho prior to the release of the **heat of liquefaction**. When that heat is released the new temperature will be that of incandescent lava. This may seem strange that so much heart energy can first be in storage at lower temperature, and then released by a change of state, all in the confined space of a lava breeder. But there is the geological evidence!

Assumption: There is little change in temperature across the Moho from oceanic crust to the material below the Moho. (This assumption is weak. There could be a temperature step across the Moho.)

Assumption: Faulting (transform faults) under the mid-Atlantic ridge extend all the way down to the Moho.

In order to fault, the temperature must be lower than the congealing temperature of basalt. Therefore, the temperature just below the Moho around a lava breeder is also less than the congealing temperature of oceanic basalt (1400 Degrees Fahrenheit). Incandescent lava, quickly moving up a heated conduit from the lava breeder to the sea bottom, probably loses little heat and therefore holds its temperature.

The temperature of oceanic basalt lava (erupting from a lava breeder) is measured on Hawaii and Iceland. According to Google, that temperature is 2100 degrees Fahrenheit.

C. MEASURING THE HEAT OF LIQUEFACTION

Three **assumptions**: 1) the congealing temperature of basalt (1400 degrees Fahrenheit) does not change with depth of burial; 2) the temperature of lava exiting lava breeders is 2100 degrees Fahrenheit, and 3) oceanic crust just above the Moho has enough strength to support differences in vertical rock frame pressure (and thus enable lava breeders).

The temperature of the material below the Moho entering a lava breeder is less than congealing temperature, say, 1300 degrees or less. Releasing the **heat of liquefaction** in the lava breeder increases the temperature by 800+ degrees, to 2100 degrees, in the material after the release of heat. That material transitions to incandescent lava. The head of the lava is somewhat less than lithostatic, because it is being released from near lithostatic into an environment of lower head above the Moho.

The 800+ degree jump in temperature is major! Therefore the high-energy "soup" infusing the mat in the material below the Moho is powerful, with lots of "other energy" contributing to the stored power! No wonder it takes millions of years for subducting basalt around the Pacific to acquire the energy needed to transition to the material below the Moho. Outer layer by outer layer, keeping the conditions of transition constant, it takes lots of time.

In today's concern for the perils of junk food, calorie content is in the limelight. Well, calories are no more than energy units. How much energy does it take to heat a fixed amount of water one degree?

In the case at hand of basalt melted in a lava breeder, estimating the annual **heat of liquefaction** released along a mile of a spreading center should be straightforward. Start with the increase of temperature (800+ degrees) and multiply by: 1) the heat needed to raise the temperature of a specific volume of basalt one degree, and 2) the volume of the "new" basalt generated that year in that mile of spreading center. The "new" volume would be the thickness of the basalt beneath the sea bottom, times a mile, times the distance the plates moved apart in a year.

Annually, a great deal of heat must leak up from the mid-Atlantic ridge into the cold water above. Measuring the amount of heat leaking up would be difficult, but a reasonable estimate should be possible.

The amount of heat lost to the Atlantic annually should equal the sum of 1) the amount of heat released annually in the spreading centers below, plus 2) a proportionate share of the heat leaking up from the earth's core.

D. HAWAIIAN AND ICELANDIC BASALT TEMPERATURES

It is interesting that the temperatures of "fresh" basalt lava in Hawaii and Iceland are about the same, 2100 degrees Fahrenheit.

In **Moho Motion** it is proposed that during the northern progression of the mounding of the Moho under the embryonic mid-Atlantic ridge, lava breeders leapfrogged north. Icelandic lava, according to **Moho Motion,** is released at the depth of oceanic Moho **next** to the island. So Icelandic lava is bred shallow, perhaps 7 miles sub-sea. Some of the lava does leak up into and onto the island. Most of the lava congeals after leaving spreading centers on the west and east flanks of Iceland.

On the other hand, Hawaiian lava presumably starts deeper, below the locally depressed Moho there. Hawaiian lava feeding Mauna Kea must have had enough head to be lifted up in the feeder tube to an elevation of almost three miles above sea level, and six miles above the regional abyssal plains near Hawaii. At the elevation of the source of the lava (lava breeder) deep below Hawaii, the pressure in the lava leaving the breeder must have just balanced the pressure in the surrounding material there below the Moho. From the pressure-depth graph in Figure 24, this lava breeder is estimated to have been at a depth of 13 miles subsea.

C. James Blom

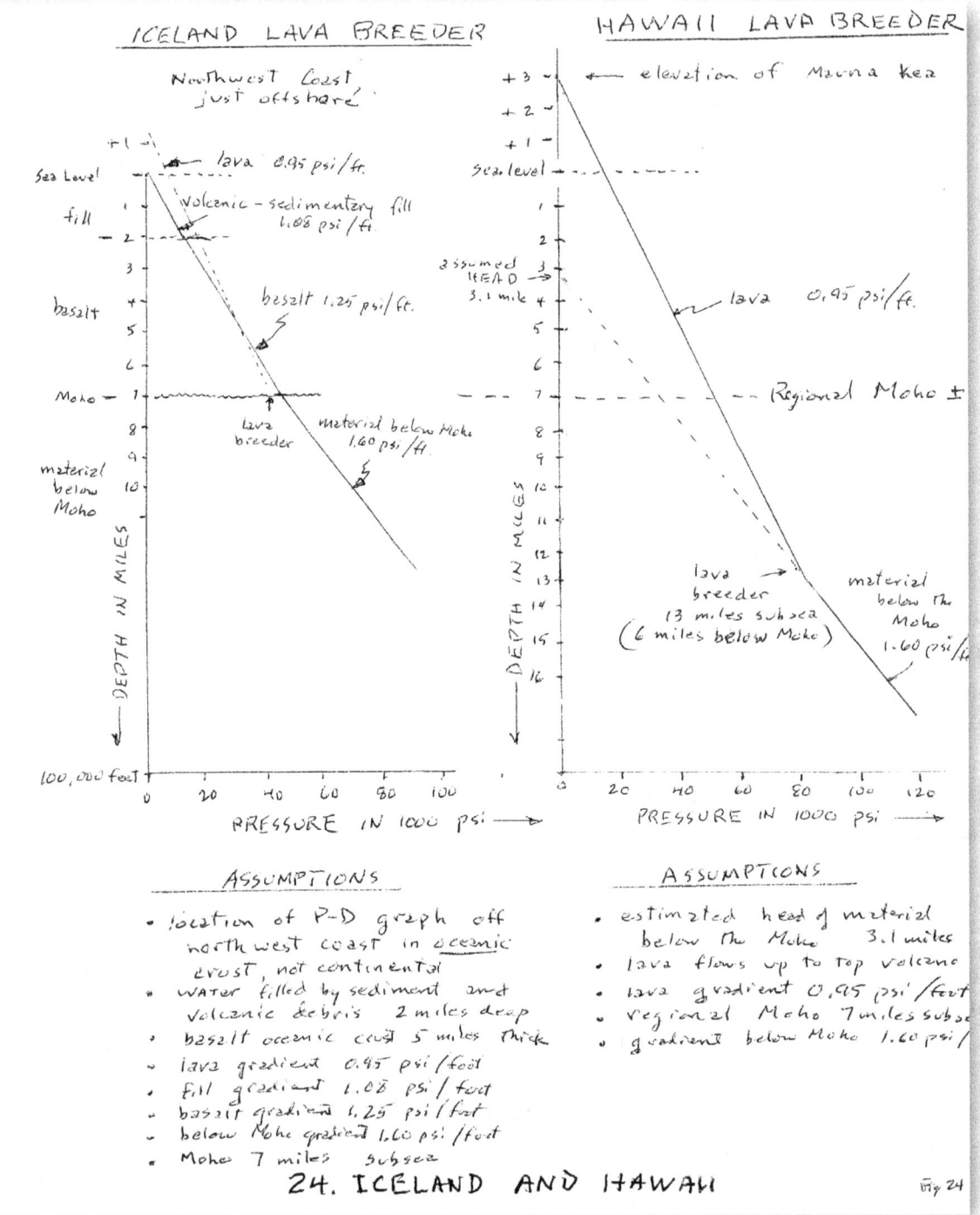

Figure 24. Comparison of the depths to the lava breeders alongside Iceland (7 miles subsea) and below Hawaii (13 miles subsea), as estimated using pressure-depth graphs.

In the Iceland graph the assumptions are 1) onshore lava reaches a height of a little over one mile, 2) the oceanic Moho is at a depth of 7 miles subsea and 3) the pressure gradients are 1.25 psi/foot for solid basalt and 0.95 psi/foot for lava.

In the Hawaiian case the assumptions are 1) The head of the material below the Moho is 3.1 miles subsea, the same as under the abyssal plains near the mid-Atlantic ridge and 2) the gradients of 1.6 psi/foot for the material below the Moho and 0.95 psi/foot for lava. Start with known point(s), add assumptions, and presto, the answer (13 miles subsea) is where the lines cross on the pressure-depth graph.

The lava breeder under Hawaii plots, on the pressure-depth graph, far **below** the estimated Moho depth. Six miles below. But what triggered the lava breeder so far below the Moho?? The Moho is probably depressed under Hawaii by the load of the volcano. Under the mid-Atlantic spreading center, lava breeders are sourced by local pressure reduction **immediately below** the Moho. Under Hawaii the lava breeder is **several miles below** the Moho.

Getting back to the temperature of the two lavas, the point is that both basalt temperatures are reportedly about the same (2100 degrees Fahrenheit) despite the 6 miles difference in depth of the lava breeder. The **conclusion** is that the 2100 degree temperature results from the release of the **heat of liquefaction**, and has nothing to do with the depth of a lava breeder.

An interesting sidelight is that the pressure in the rising lava in a feeder tube "should" break out the side of the volcano and form a side cone. Rising lava with a gradient of 0.95 psi/foot, in the Iceland pressure-depth graph, exceeds the pressure in the surrounding basalt above 4 miles subsea. The fact that lava doesn't break out could mean that the feeder tube is somehow structurally strengthened, and lava remains corralled.

E. HAWAII'S PREDECESSOR VOLCANOS

Hawaii is the latest in a string of predecessor volcanos, of which the submarine remnants are strung out on the sea bottom for hundreds of miles northwest of Kauai on the abyssal plain. About a hundred million years ago Hawaii's submarine linear track of ancient volcanos changed directions. A kink in the alignment! It would appear that regional constraint on the Pacific plate changed, and thereafter plate motion went in a different direction over the stationary material below the Moho. (And over the stationary lava breeder.)

Below Hawaii could there be another type of trigger mechanism for a lava breeder causing a persistent, local site of unstable low pressure in the material below the Moho? This mystery-trigger operated for more than a couple hundred million years in the stationary material below the moving crust. Let your imagination run wild. What about a slow speed rotation around a vertical axis in the material below the Moho? For an analogy think of a tornado. The vertical vortex could have lower head, sufficient to intermittently trigger a lava breeder there. Why would the breeder be 13 miles subsea, instead of 10 or 16?? Perhaps the vortex of the cyclonic disturbance has its lowest head a certain distance below the Moho.

Nature does funny things. The Great Red Spot on Jupiter, an enormous cyclonic storm, has been observed continually since it was first seen in 1831. Given nature's quirks, perhaps a very slow speed cyclonic rotation in the material below the Moho, that has lasted a couple hundred million years, is not out of question. It would solve a conundrum here.

Under Hawaii at 13 miles subsea, pressure in the material below the Moho is almost 80,000 psi. If the pressure reduction in the vortex at breeder depth were only 100 psi, perhaps that would be sufficient to trigger a lava breeder there. Perhaps the vortex trigger would be activated only in a particular phase of the moon, or annually when closest to the sun.

Tut, tut! We are stretching here!

F. TEMPERATURE OF TRANSITION FROM BASALT TO MATERIAL SUB-MOHO

As long as we are dealing with subsurface temperature estimates, let's make a guess about the subducting oceanic crust around the edges of the Pacific. **Moho Motion** proposes that the cool slabs of oceanic crust slowly follow down cooled channels hundreds of miles in the material below the Moho. Inch by inch annually. That's subduction. The top and especially the bottom of the slabs are heated by radiant heat coming up from the core, and subjected to electrical currents. At some specific temperature, or narrow range of temperatures, the outside of the descending basalt slab slowly transforms into the stable material below the Moho. It's a change of state: basalt to material below the Moho.

At what temperature?? In the preceding section the estimate of the temperature of the material below the Moho is 1300 (or less) degrees Fahrenheit when it is stable and is about to release the **heat of liquefaction**. If that is the temperature when it releases the heat, isn't that a good estimate for the temperature when oceanic basalt would accept rejuvenating heat? By analogy, boiling water at atmospheric pressure gives off steam at 212 degrees Fahrenheit, and when steam is cooled it condenses at atmospheric pressure into water at that same temperature.

So the ball park estimate of the temperature necessary for basalt to transform state, upon heating, from basalt to the material below the Moho, is 1300 degrees Fahrenheit, or less. Because heat is absorbed only on the surface of the descending slab of basalt, and because the slab descends at the speed of plate spreading, the heated side of the slab will remain at nearly constant temperature for millions of years until, layer by layer, the last of the basalt transforms. During the heating process, electric currents may add to the energy needed to increase the density of the mat of "soup" and crystals, comprising the material below the Moho.

G. HORIZONTAL STRESSES IN BUOYED BASALT

We have focused on the **vertical** pressure exerted by a layer of water saturated rock on its support (the next layer down). This vertical pressure is the sum of water pressure plus the vertical component of the buoyed stress in the framework of the rock. Now let's consider the **horizontal** stress. In a placid environment like the abyssal plains on either side of the mid-Atlantic ridge, the horizontal stress in the rock frame should be the same in all compass directions. In the rock frame the horizontal stress is approximated as the vertical stress in the rock multiplied by twice the Poisson's ratio. Assuming the Poisson's ratio is 0.25, the horizontal stress in all directions in the rock frame would be 50% of the buoyed vertical stress.

H. PRINCIPAL PRESSURES

As noted earlier, engineers deal with stress in an elastic solid by dividing it into the three principal stresses at right angles to each other. Often in geology, fluid pressures as well as the principal stresses in rock are combined. For example, lava (fluid) intrudes rock (solid rock plus water in the pores) along a zone of weakness by pushing its way in. " Pushing its way in" means the lava must have enough pressure to shove the two sides of crack apart. Lava pressure overcomes the combined horizontal buoyed rock frame stress plus water pressure.

This introduces the concept of **principal pressures**. Solids have principal stresses in three dimensions. Fluid pressure at a point is the same in all directions. When both pressure and a solid principal stress are considered together, "principal **pressure**" is an appropriate designation. In a submerged solid there are three principal pressures, oriented parallel with the principal stresses in the buoyed rock frame.

As a mental crutch, insert an imaginary diaphragm (oriented vertically) at a reference depth in the basalt below the mid-Atlantic ridge. In order to inflate the diaphragm in the horizontal direction, the pressure applied must be equal to the total horizontal pressure perpendicular to the diaphragm (buoyed rock frame stress in that direction, plus liquid pressure). If the vertical diaphragm is oriented north-south then the principal pressure needed to inflate will be the minimum amount -- in the direction of the pull-apart.

In Figure 25 dry sandstone has density about 2.5, equal to a pressure gradient of 1.08 psi per foot of depth, or 1080 psi at 1000 feet depth. The horizontal stress in all compass directions in the dry sandstone is 540 psi, thanks to the assumed Poisson's ratio of 0.25.

Next fill the sandstone with a continuous column of fresh water. The new vertical pressure at 1000 feet depth is again 1080psi, comprising 430 psi from the fresh water column, and, by subtraction (1080 – 430 = 650), the buoyed sandstone's share of vertical pressure is 650 psi.

The horizontal stress in the buoyed rock frame of the sandstone is added to the graph. Note that the horizontal stress in the buoyed rock, 325 psi, **is 50% of the buoyed sandstone**, not 50% of the dry sandstone

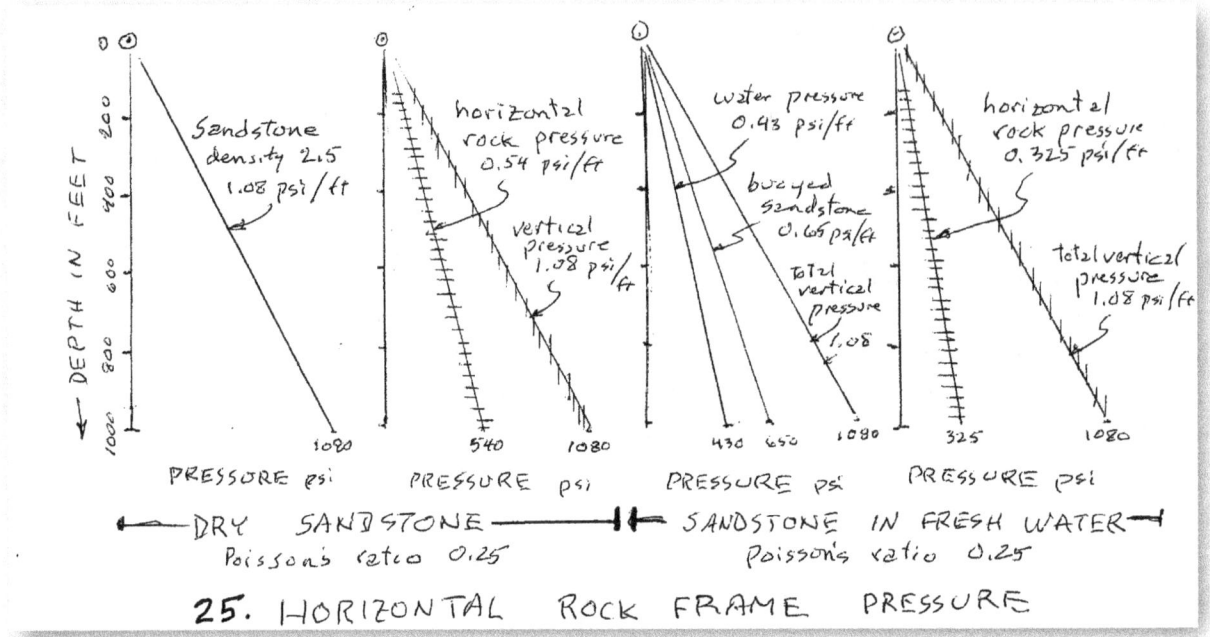

Figure 25. Comparing horizontal rock frame pressure in dry sandstone, and in the same sandstone saturated with fresh water. Assumes Poisson's ratio is 0.25, meaning horizontal rock frame stress (pressure) is half the vertical rock frame stress. Note the difference in horizontal stress in the frame of the rock: buoyed 325 psi, vs 540 psi dry.

Getting back to the horizontal stress in rock buoyed up by sea water, under the abyssal plains flanking the mid-Atlantic ridge, the horizontal stress is essentially the same in all directions. Evidence for this is undisturbed oceanic dust and lack of earthquakes there.

I. MID–ATLANTIC RIDGE PULL-APART

At the crest of the mid-Atlantic ridge there is ample evidence of a pull-apart and the creation of new oceanic crust. Here stress is least in the direction of the pull-apart. The basalt is buoyed by sea water as is evidenced by hot flows of mineral-rich sea water observed by diving craft. We have noted that this condition has existed long enough to support the evolution of strange creatures feeding on the chemicals in the hot water, so it has been a constant condition. **Assumption**: superheated sea water extends down to the Moho, and it cools the lava, causing the rising lava to congeal.

In the crest of the mid-Atlantic ridge, the buoyed stress in horizontal directions in oceanic crust differs greatly. In the direction of the west-east pull-apart, stress is least. But this stress remains **compressional** (except locally near the sea bottom). This may not seem logical because the crust is being separated in the "pull-apart" direction. So if it is pulling apart, it 'should" 'be in tensional stress (as opposed to compressional).

But let's consider the effect of Poisson's ratio. If the material in the pull-apart were cork, indeed the separation between plates would be a vertical crack, held together at the base by the tensile strength of the cork. But the material isn't cork, it is basalt, and its Poisson's ratio is assumed to be 0.25. This means 50 % of the vertical stress in the rock frame is distributed horizontally in the north-south direction (intermediate stress). But in the west and east directions of movement of the two plates, stress will be much less, albeit compressional.

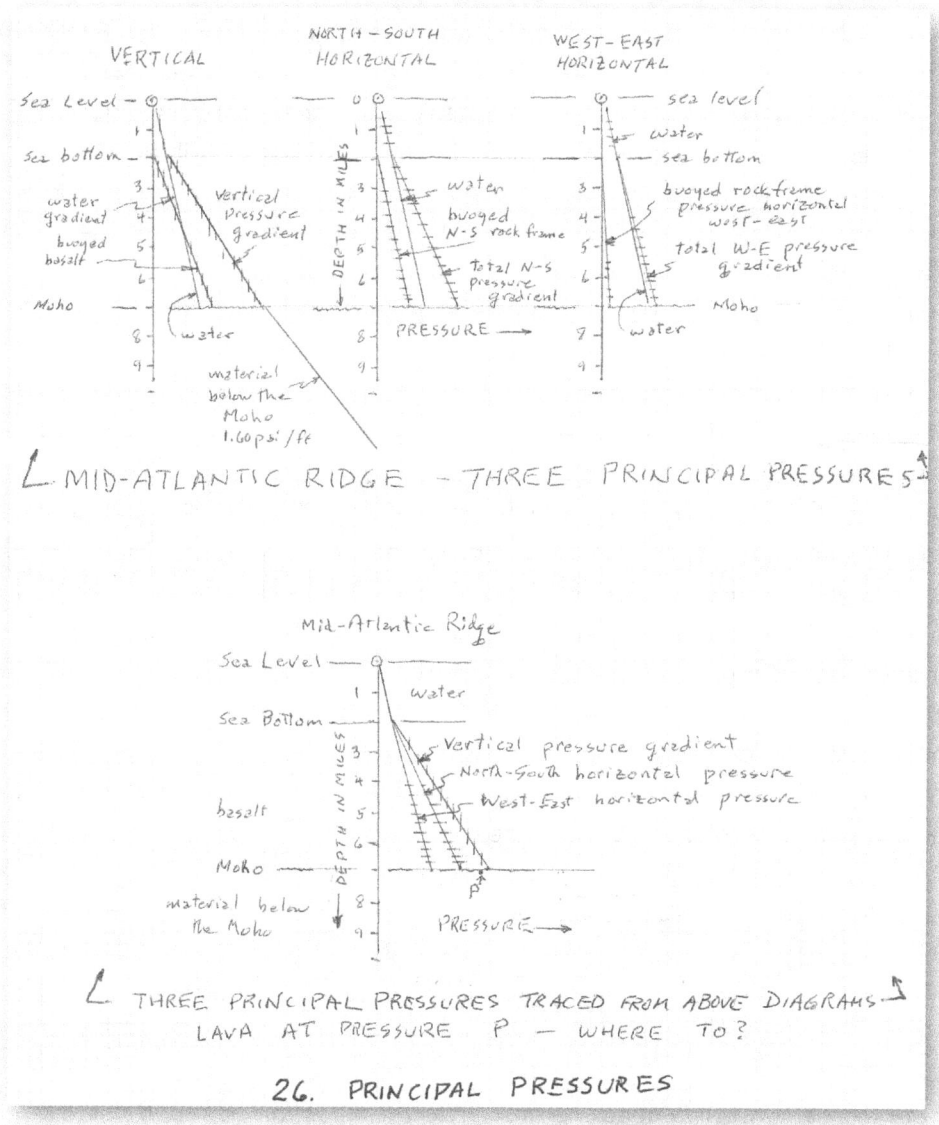

Figure 26. Upper profiles. Three principal pressure profiles at the crest of the mid-Atlantic ridge. 1) vertical pressure on the Moho, 2) horizontal north-south pressure, 3) horizontal west-east pressure in the direction of the pull-apart.

Lower graph. The three principal pressure profiles, traced from the above graphs, show the stresses faced by lava released just below the Moho under the crest of the mid-Atlantic ridge. What would a lava, with pressure P, do to buoyantly escape?

Breather

Here is a little breather for the tortured audience. Kick your imagination into high gear. You are now on the spot. You are a large load of lazy lava. You are at pressure point P on Figure 26. What you have going for you is pressure, lots of pressure. And you are hot! It's decision time. Now quickly, what are you going to do?? Escape you will. But by what route? And why?

The pressure-depth graphs in Figure 26 need a little explaining. The objective of the diagram is to highlight pressure distribution controlling lava flow up from just below the Moho into a mid-Atlantic spreading center. All the cartoon's gradient lines are taken by light table from Figures 18, 19 and 20. In each case water pressure is added to the buoyed rock frame pressure to get the total pressure. Water is assumed to exist from the surface to the base of the basalt. Total vertical pressure gradient for **buoyed basalt plus water** is the same as the gradient for dry basalt (water cancels out). The assumed Poisson's ratio of 0.25 applies to the rock frame pressure of buoyed (as well as dry) basalt.

The bottom diagram has the same three pressure gradients, copied by light table. The meaning is hard to visualize. Look to the corner of the room you are in to see walls and floor meeting at right angles. Pressure on the north wall comes from the north and is balanced by pressure from the south. Pressing against the west wall is the modest pressure in the pull-apart zone, and it is balanced too. North-south direction pressure is strong, but west-east pressure is very weak (but compressional).

Above the Moho, if the buoyed stress in the rock frame would become extensional (forming open cracks) this would be an open invitation to buoyant lava to invade. Lazy lava would already have forced its way into open cracks. **Conclusion: open** tensional cracks/joints don't exist in the mid-Atlantic ridge except locally in the cold brittle basalt just below the sea floor. At shallow depths below the sea floor some lava, having risen five miles, may not fill the surface crack at the spreading center because, following its circuitous ascent, it has bled off nearby onto the sea floor. Later the surface cleft will be filled by lava from another lava breeder.

As lava intrudes and congeals in the lower part of the basalt beneath the mid-Atlantic ridge, the west-east pressure in that buoyed section is increased. This decreases the amount of bulge in that direction. With less bulge (from vertical loading), stress in the west-east direction increases. Eventually, at some distance from the spreading center, the cumulative effect of all the intrusions of lava into the fractures and joints causes the horizontal stress in the solidified basalt to be the same in all horizontal directions. This condition of "healed" cracks/joints in the basalt is the norm under the abyssal plains.

Reprieve

OK, so you were left hanging. You need the answer to the question posed in italics a few paragraphs back. You are not comfortable in your new role as a large load of lazy lava. Being lazy, your instinct is to head for the easiest way out. Correct! Starting at pressure P your correct choice is the lowest pressure exit. That would be where the cracks and joints oriented north-south are held shut by the least pressure. Some of your lava would probably rise up hundreds of feet in the cooled cracks. There it would congeal!

Congratulations, you are a Dad/Mom!! Your tyke is so cute! The little shaver looks a little like the bellow of an accordion when the accordionist swings his left arm wide for more air. Oh, by the way, the accordionist is lying down on his right side. He scooches around so his head points north. Now when he raises his left arm, the creases in the bellow are horizontal and point north-south. That looks sort of like your kid, except your offspring is much, much larger, he is irregular, and he is solid basalt.

Then, with most of your pressure still intact, you would find another crack trending north-south. This time your lava might rise further before congealing. Several cracks later your lava might rise to the ocean floor and spread out there.

When your pressure poops out, you are a goner. Another nearby lava breeder will take over. Further away from the spreading center, all those "accordion bellows" coalesce into solid basalt, and there are no more cracks to fill, because the horizontal stress in the rock frame is the same in all directions.

❖　❖　❖

Eighteen

Oil Pools,
Mt. St. Helens and
the mid-Atlantic
Ridge

A. ENTRAPMENT OF ENERGY

The title of this chapter might sound like the author slipped a cog, and is mixing apples, oranges and grapefruit. But there is method in this apparent madness. All three topics describe nature's trapping of fluids in porous rocks. The common thread is "**cap rocks**", or the trapping mechanism. The other common thread is the **method** of describing and understanding how fluids are trapped in rocks, namely pressure-depth graphs.

Cut me a little slack here. In order to understand the Mt. St. Helens eruption you will need a little boring background, best introduced through petroleum reservoirs. Hang in there, and then we will get on with the exciting stuff.

The mantra of petroleum explorationists up until recently has been: "Source Rock, Reservoir rock, Cap Rock". The logic is that petroleum is sourced from a high concentration of organic matter entombed in rock at the time of deposition. On the other hand vegetable matter transitions to coal, but let's stick with petroleum. Rich organic matter very slowly changes into petroleum depending on depth of burial and higher temperature. Oil or natural gas then migrates from the source rock into nearby porous reservoir rock. The cap rock over the structure makes it impossible for the buoyant petroleum to escape, despite its buoyancy. Petroleum is trapped by a cap through which the buoyant petroleum cannot penetrate.

Since the turn of the 21st century, American entrepreneurs devised methods to profitably produce petroleum directly from source rocks which are so impermeable that petroleum remains trapped there. In this new technology the source rock is fractured, so that source rock doubles as the reservoir rock.

"Notorious new" fracking techniques, in service for decades without problems, release petroleum directly from the source rock into the wellbore, bypassing the need for a reservoir rock.

At Mount Saint Helens, Washington, rain water, keeping the water table close up under the land surface, trapped steam in areas of active volcanism.

Similarly, under the mid-Atlantic ridge, two miles of dense cold sea water plus the water in porous basalt below, is a (leaky) trapping condition for the less-dense hotter water below. Sea water in the pores of the basalt is heated by radiant heating from the release of the **heat of liquefaction.**

B. OIL AND GAS FIELDS
Buoyant fluids like petroleum will rise in reservoir rock until buoyant movement upward is stopped by conditions in the overlying rock. At the oil-water contact the pressure of both fluids is the same. The surrounding water pressure determines the pressure in the trapped petroleum. Above the level of the oil/water interface the buoyant petroleum is at higher pressure than the intimately surrounding water. Moving higher in the reservoir, buoyancy of the oil forces the water coating to become thinner. Thinner water means thicker petroleum, thereby concentrating more petroleum in the pores. Petroleum engineers, computing field reserves, pull out all the stops getting accurate answers here.

The increasing concentration of buoyant petroleum in the pores of the reservoir continues upwards, a) until the overlying cap rock becomes impermeable to the buoyant petroleum, b) until the nature of the reservoir rock changes adversely, c) until the head of water in the overlying cap rock is greater than the head in the water on the surfaces of reservoir pores, or d) until the film of water coating the pore walls cannot be squeezed any thinner (irreducible water saturation).

Irreducible water saturation means that the thin layer of water plastered on the walls of the pores obeys molecular forces rather than hydraulic (buoyancy). Within the pores in the reservoir, molecules of water interact with molecules in the walls of the pore. These attractions are much stronger than hydraulic forces. Molecular forces trump buoyancy pressures when the film of water gets thin.

C. BUOYANT PETROLEUM GETS ITS
PRESSURE FROM WATER
Note that at the original condition within the petroleum reservoir, before any petroleum is produced, every petroleum accumulation is static, i.e. it is not moving in the reservoir. There may be a gradient of water head across the reservoir, which indicates that the water is moving through the reservoir, albeit slowly, but meanwhile pressure in the virgin petroleum accumulation remains static.

For every buoyant petroleum accumulation, the original reservoir pressure is in equilibrium with the water pressure at the petroleum/water contact.

In figure 27 the column of light oil has higher pressure than in the enclosing water, and the pressure difference at any particular depth is the buoyancy pressure there. The cap rock above the oil column successfully resists the buoyancy (or the oil would leak out the top).

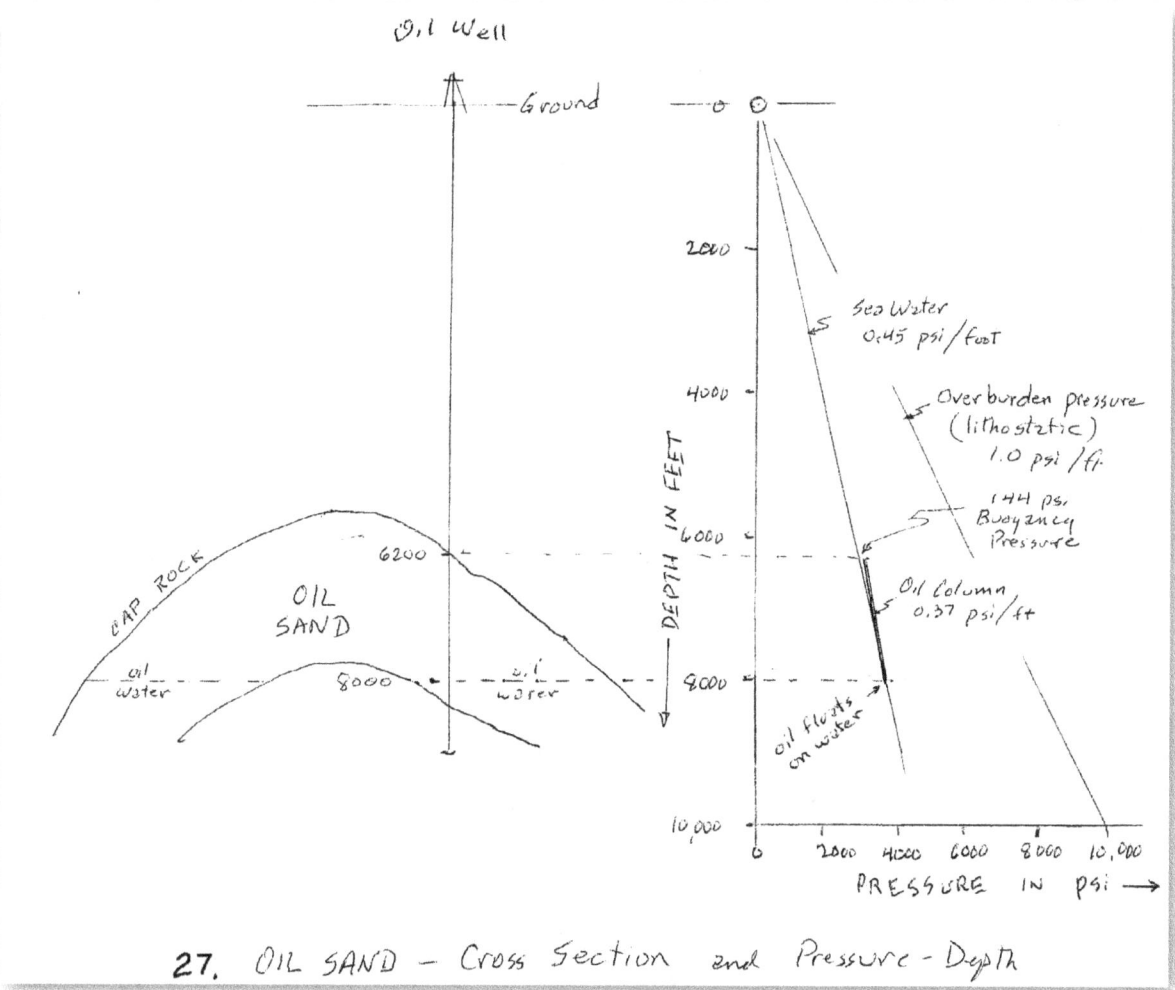

Figure 27. Oil Well with 1800 foot oil column. Buoyancy pressure at the top of the oil zone in the well is calculated subtracting oil gradient (0.37 psi/foot) from water gradient (0.45 psi/foot) equals 0.08 psi/foot times 1800 feet, or 144 psi. The cap rock doesn't leak despite 144 psi buoyancy pressure.

The oil industry uses it's very own scale for defining density of oil. In this artificial scale the units are "degrees of gravity", or just "gravity". Heavy oil with the same density as water is 10 gravity. Think Athabasca tar sands in Canada. Light oil with density of 0.7 is 35 gravity (0.37 psi/foot). (Light oil with low sulfur content is favored by refineries.)

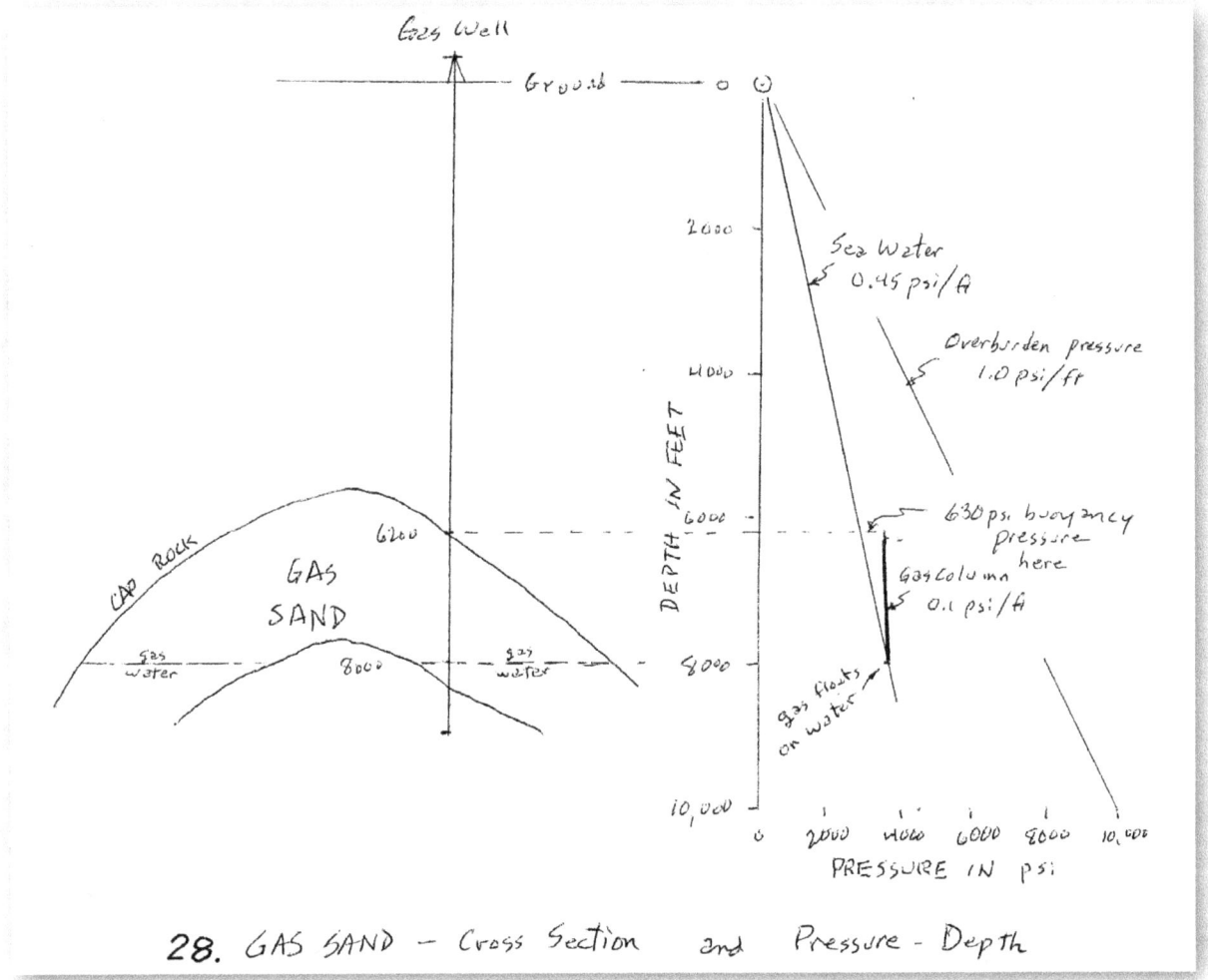

Figure 28 Gas Well with 1800 foot gas column. Calculation as in figure 27 results in 630 psi buoyancy pressure at the top of the gas zone in the well. The cap rock is probably over-pressured (higher head) in order to withstand that much buoyancy without leaking.

Natural gas density is very light, but it varies with pressure, temperature and the content of gas liquids. If ethane, propane, butane and yet heavier molecules are mixed in with the main component of natural gas, methane (CH_4), density of the natural gas will increase. Cap rocks for gas accumulations resist much more buoyancy than for oil, because the gas is lighter. The original reservoir pressure in each pool of trapped gas varies vertically only as a function of the density of the gas. Again, **the original reservoir pressure is at equilibrium with the water pressure at the gas-water contact.**

In a petroleum reservoir with very small pores it may take a few psi of extra buoyancy to overcome the entry pressure of the petroleum into the pores. But the situation remains unchanged: the pressure of the water at the theoretical water/petroleum contact determines the petroleum pressure in the reservoir. Above the water/petroleum contact the petroleum pressure varies according to the density of the petroleum.

Now let's consider a simple petroleum reservoir in three dimensions (Figure 29). The anticline has four way closure, meaning the crest is higher, so buoyant petroleum in a reservoir under the cap cannot escape in any direction.. Below the cap rock is a reservoir rock with porosity for the petroleum. Nearby under the cap rock is a source rock from which petroleum is expelled. The buoyant oil migrates into the reservoir rock.

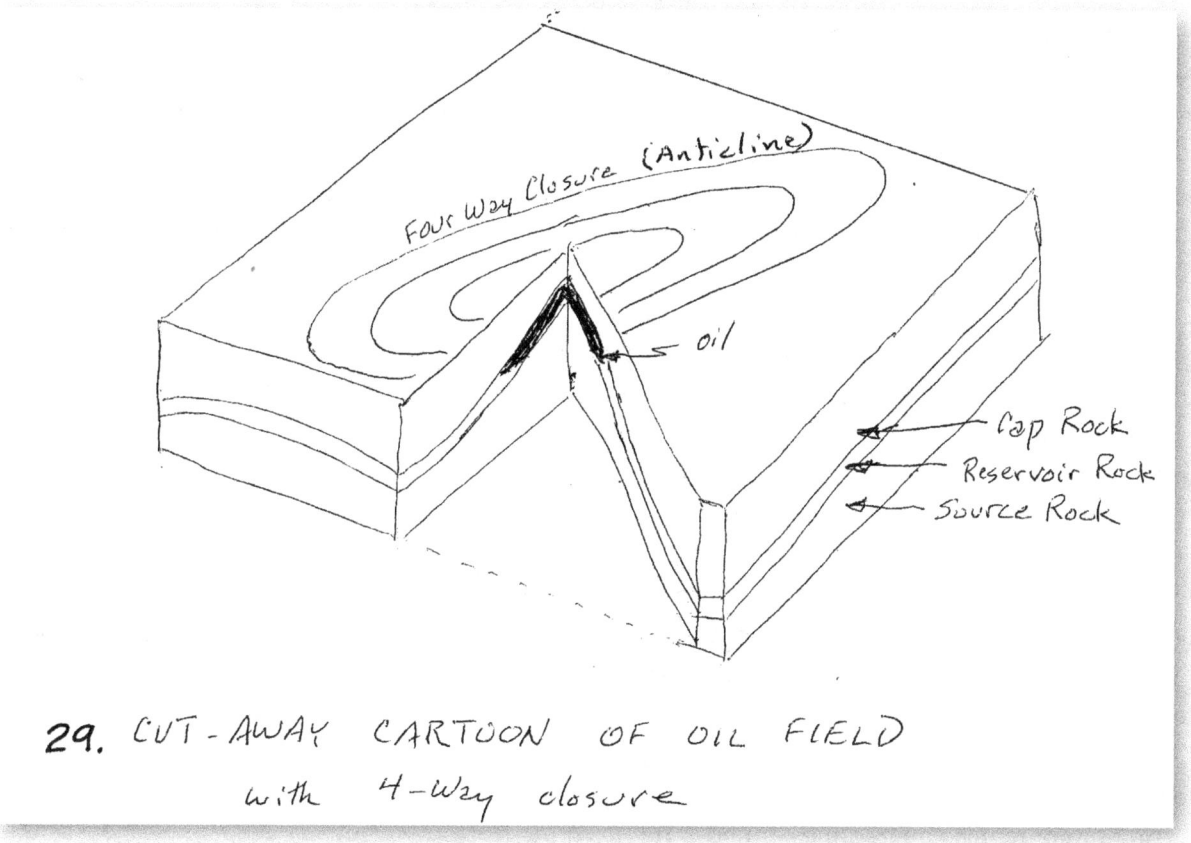

Figure 29. Petroleum Field in an unconvincing 3-dimensional cartoon.

In a petroleum pool the water pressure gradient is assumed to be continuous at 0.45 psi per foot, but it cannot be measured directly. Only the petroleum is mobile, so only its pressure is measurable.

A large percentage of petroleum reservoirs had original pressure either greater, or less than hydrostatic. Most of the recently-developed "unusual" oil and gas accumulations in shale are at original pressures higher than a column of water to that depth ("over-pressured").

D. 1980 ERUPTION MOUNT SAINT HELENS, WASHINGTON

On May 18, 1980 a small spotter airplane was flying over Mt. St. Helens, monitoring volcanic activity. For the past months the mountain had all the symptoms of a volcano about to go into action. The northern flank of the mountain side was slowly bulging out, and daily earthquakes were recorded.

Looking down, the pilot saw an angry jet of dark debris exploding out from the mountain side. The pilot turned tail and dove steeply, successfully outrunning whatever was behind him.

Gary Rosenquist was eleven miles away with his camera set up on its tripod. There was a 5.2 earthquake under Mt. St. Helens. Seconds later Gary saw a huge landslide starting down the north side of Mt. St. Helens. He rose to the occasion and commenced snapping photos at reasonably equal intervals. Several seconds later the dark jet of steam and reworked ash exploded forth out of the upper reach of the avalanche. The ferocity of ensuing explosions reduced the elevation of Mt, Saint Helens by more than 1000 feet.

Within ten minutes the cloud of hot ash and dust rose to 80,000 feet. Volcanic ash rained down over Washington and 11 other states. Gas and ash continued belching with diminishing force for days, months and years.

For an excellent account, minute by minute, of Mt. Saint Helens on May 18, 1980, consult the Wikipedia entry on your computer.

OK, what's the connection to cap rocks?

First we must recognize that at Mt Saint Helens there were two independent fluid systems in operation: magma (subsurface lava) and water. The magma was sourced intermittently from a subducting slab of oceanic crust several tens of miles below. The water system was sourced by rain. Radiant heat from the magma heated the water. Let's focus on the water system.

Annually Mount St. Helens gets much more than 100 inches of rain, keeping the water table close under the topography.

The water below the water table acted as the cap for trapped steam. As in petroleum accumulations, the pressure in the steam was in equilibrium with the pressure in the water at the highest elevation of the water/steam contact. Steam is much less dense than the cold water below the water table. The pressure gradient within the steam depended only on the density of the steam.

Prior to the 1980 eruption, a massive new pulse of intruding magma forced its way into the root of the Mt. St. Helens, causing the northern surface of the mountain to bulge out 450 feet. This hot subterranean newcomer also heated the ground water from the bottom up.

For tens of thousands of years under Mt. St. Helens, subsurface magma heated water in the porous basalt. Heating varied depending on magma distribution below ground

As the temperature under the ground water increased, ground water was transformed to steam. Thus, nature was storing heat energy trapped in the pores in the rock.

Steam was the backbone of the industrial revolution. Steam has been studied to a fare-thee-well for much more than 100 years. Google "steam tables" and you will find two kinds. One is for gourmet chefs to keep their delicacies piping hot. You want the other. The data for Figure 30 were extracted from the Googled **Properties of Saturated Steam – Imperial Units** from **The Engineering Tool Box.**

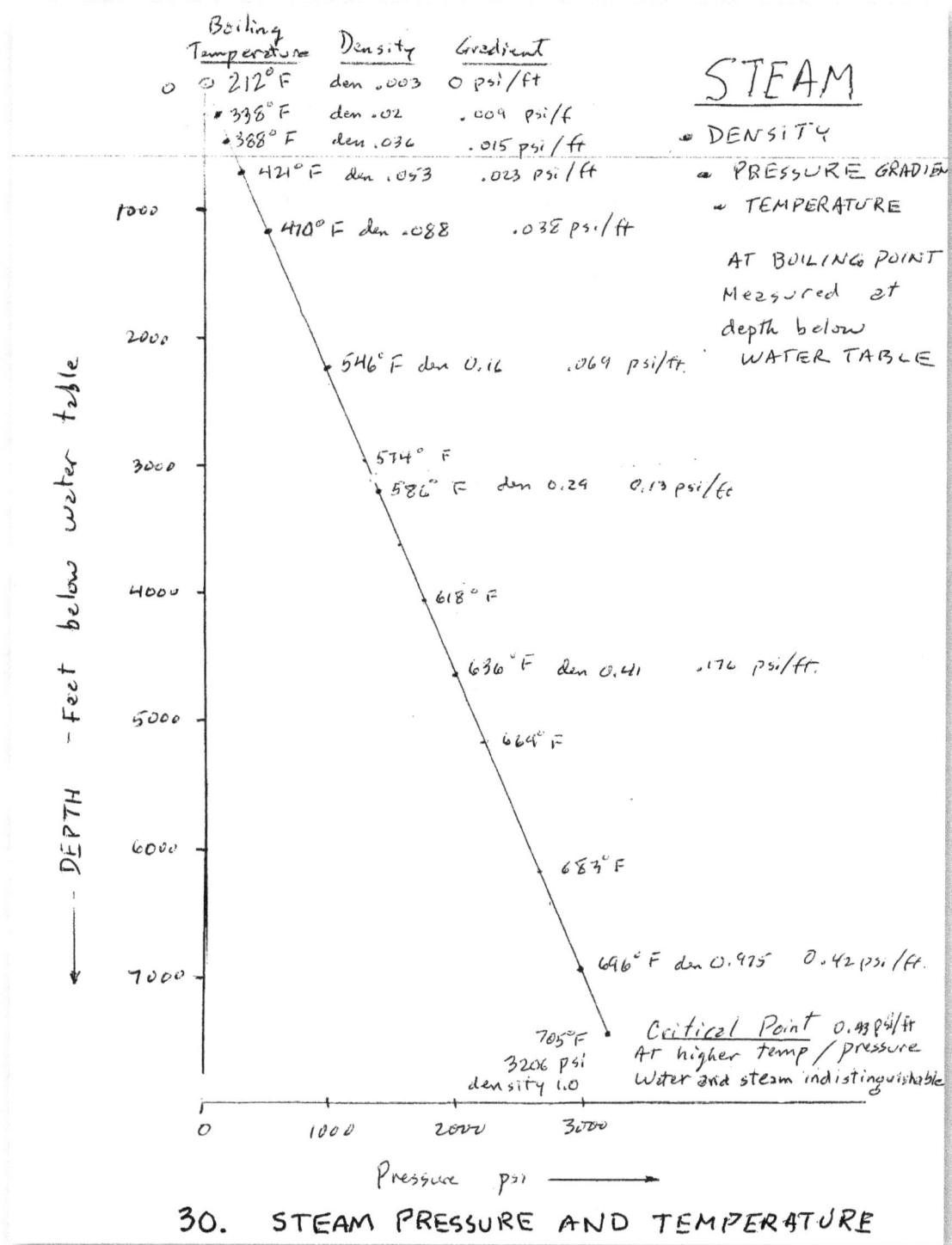

Figure 30. Steam boiling temperatures, densities and pressure gradients for depth below the water table.

It takes longer to boil potatoes at high elevation, because the temperature of boiling is less than at sea level. Increasing the pressure drives the boiling point of water right up to 705 degrees Fahrenheit. Then something strange happens. At higher temperatures, water and steam are one and the same. They are indistinguishable.

This temperature and pressure is called the critical point. For fresh water this temperature is about 705 degrees Fahrenheit and the pressure in porous rocks is equivalent to a column of water about 7500 feet below the water table.

The moving landslide photographed by Gary Rosenquist on May 18, 1980 transferred potential energy (height) into kinetic energy (increasing velocity) of the moving landslide. As the landslide progressed, the speed of the slide increased. Gravity usually just holds objects in place. But in this case a significant portion of gravity accelerated the landslide downhill. The vertical rock frame pressure on what now was the base of the landslide was lessened. Pressure in the water within the slide also dropped suddenly because the water was now losing elevation as the slide was picking up speed. The landslide was so thick that the base of the landslide probably extended down into the steam trapped below the water-steam interface. Steam had been contained in the pores of the old volcanic debris by the pressure of the overburden, but now it was free! The steam was unroofed!

The viscosity of live steam is far, far less than that of water. The pressure of steam in the pore space underneath the landslide was no longer capped. The steam pressure below the sole of the landslide did not change, but the pressure which had been containing the steam dropped precipitously. The combined effect was an **UNDERGROUND STEAM EXPLOSION!!**

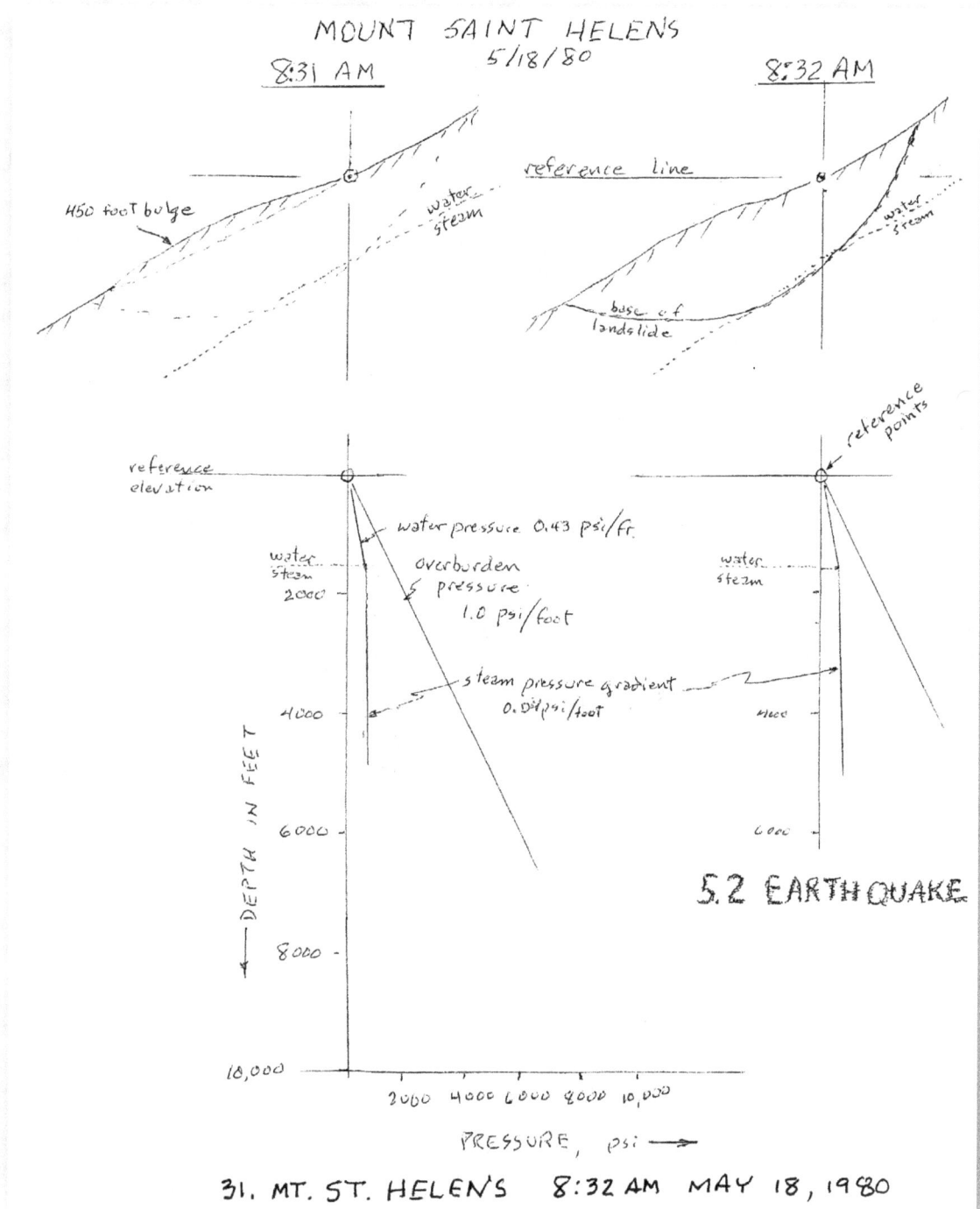

Figure 31. Mount Saint Helens, 8:31 AM and 8:32 AM, May 18, 1080. This cartoon shows the start of the landslide, which would quickly release a steam explosion. The upper cartoon is a cross section of the action, and below is a pressure-depth diagram. The circled reference point is the same for every diagram.

The steam in the pores nearest the pressure reduction (base of the landslide)l expanded rapidly. The steam explosion started with the gusher of black material from the upper part of the landslide that startled the pilot of the spotter plane. This initial gusher was captured on the motion picture of the start of the event (on your computer beam up a documentary: **Rare Film Footage of Mt. St. Helens Volcanic Eruption**, courtesy US Forest Service). The pressure holding back the steam dropped even further as the overburden was blasted away. The explosion intensified.

Steam explosions are nasty. Reports of steam locomotives exploding are chilling, … er.. scary. The strength of the volcanic debris constituting Mt. St. Helens was no match for the force of steam exploding out of the pores of the rock. The exploding steam carried massive amounts of the mountain with it. The back pressure from moving all that rock slowed the speed of the explosion.

The steam was released along a very rapidly expanding underground front. The front moved explosively back into the volcanic debris. In milliseconds, the raging turbulence of released steam increased as the area of the underground explosive front expanded. Steam inside the pores in pre-existing volcanic rocks exploded, making little ones out of big ones. Steam at 500 degrees Fahrenheit mixed with volcanic debris to form a well lubricated hot mobile density fluid. The mixture explosively welled up through the gaping hole which formed below the initial surface landslide.

As the steam explosion matured within just a minute or two, it ran short of steam, felt the increasing backpressure of debris and water, and eventually died, leaving only escaping steam and other gas. Meanwhile the debris flow, of steam mixed with rock, continued downslope at speeds up to hundreds of miles per hour, flattening forests and scalding anything alive. The gas in the debris flow was only steam, so all living creatures caught there were asphyxiated. In the above referenced film documentary, several scenes show the parallel downed tree trunks aligned across the countryside, giving a clear picture of the direction of movement of the steam/debris flow. Great clouds of escaping steam eliminated any visibility.

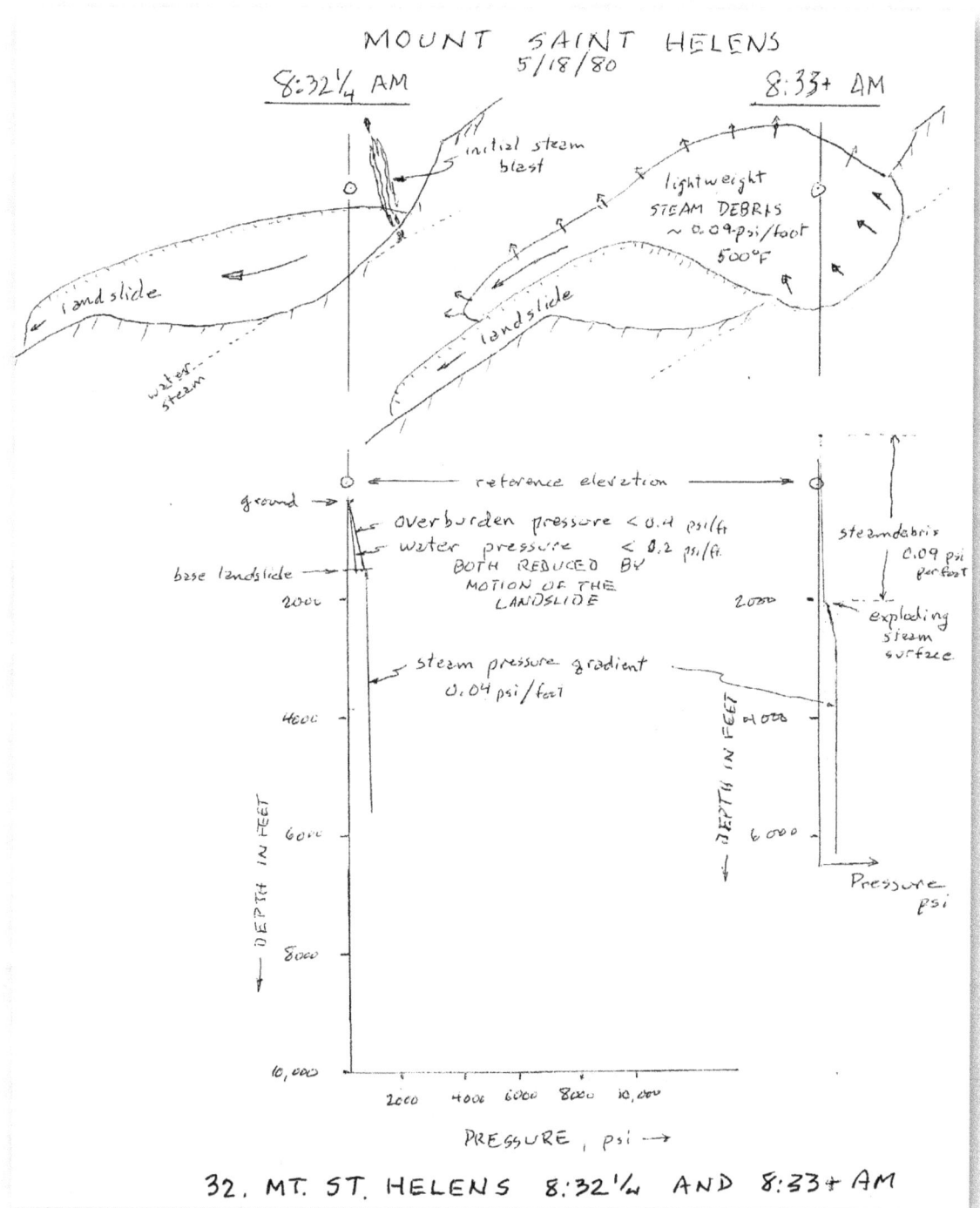

32. MT. ST. HELENS 8:32¼ AND 8:33+ AM

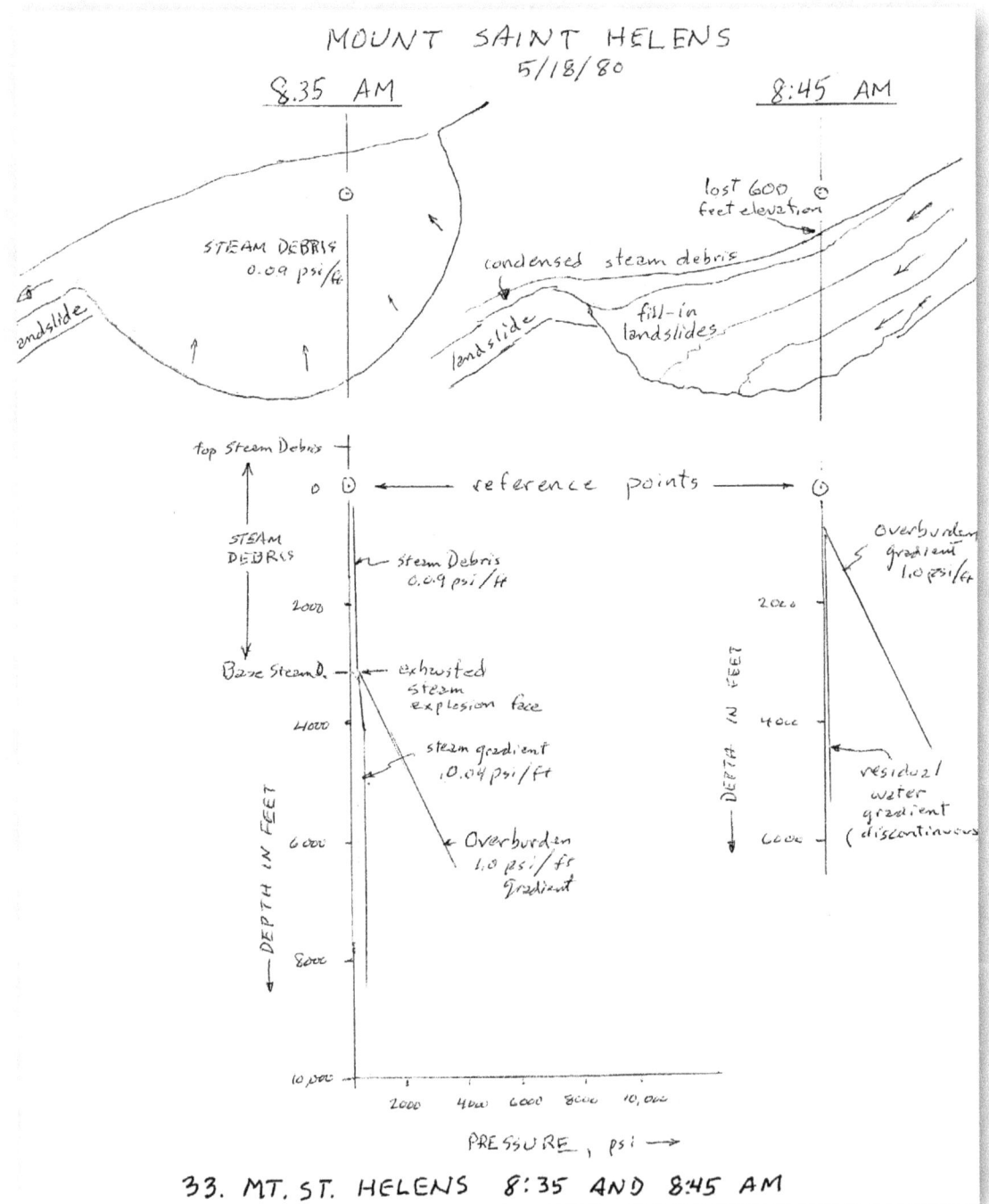

Figure 33. These two cartoons continue the events of figure 31 to 8:32 ¼ AM, 8:33 + AM, 8:35 AM and 8:45 AM on May 18, 1980. No one was standing around with a stopwatch, and these events could not have been observed anyway. So this sequence is speculative, but well founded.

Figures 32 and 33 are diagrammatic **two**-dimensional cartoons. Better would be a **three** dimensional picture identifying the shallowest "top" of the steam. The pressure in the steam at the top would be the same as the trapping water pressure at that depth. The temperature of the steam at the water/steam contact would be the boiling temperature of water at that pressure. Steam pressure elsewhere in the steam reservoir would be distributed according to the pressure gradient of the stream.

This brings up a paradox. Intuitively you would think that increasing the heating of the steam in the volcano's pores would increase the pressure and temperature. Actually, it's the opposite. Let's look at a hypothetical example. In figure 34 the same cross section of Mt. Saint Helens is "previsited" in 1900 (more or less) and in 1979, a year before the explosion.

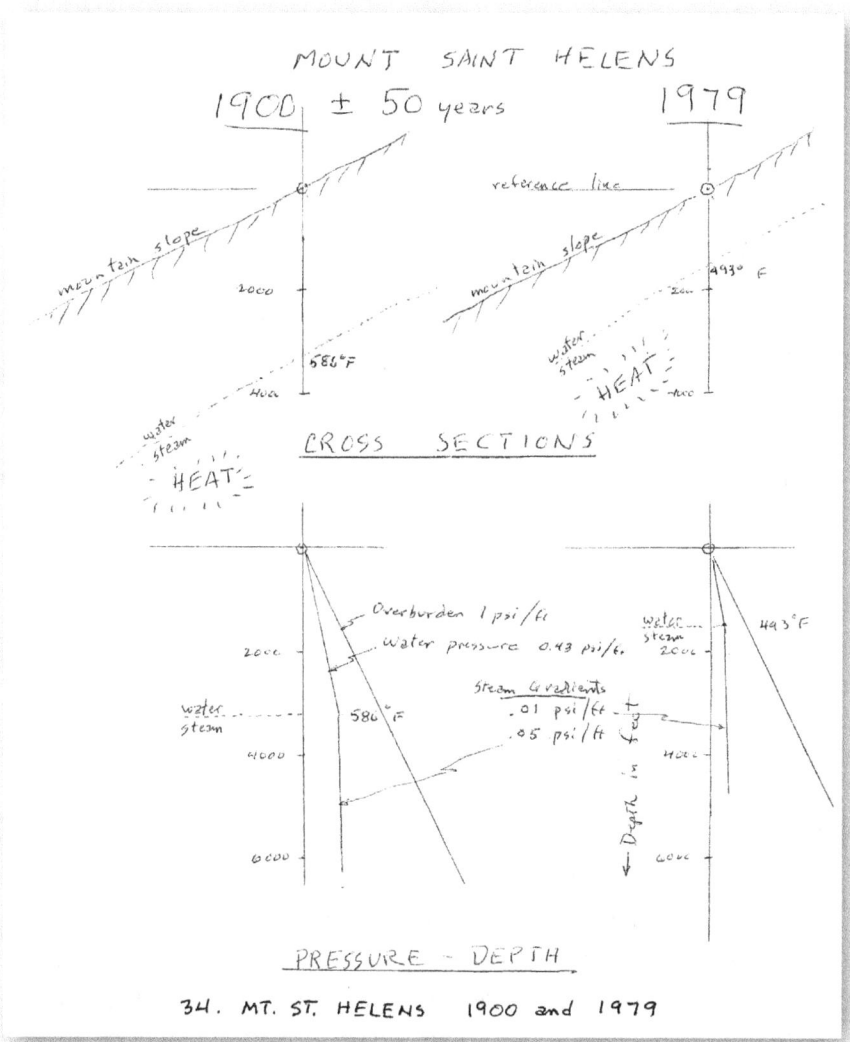

Figures 34. Two cross sections (top) and corresponding pressure-depth graphs (bottom) for the same location as in figures 31, 32 and 33 for years 1900 (plus or minus 50 years), and 1979.

In year 1900 the water steam contact is depicted at 3200 feet below the water table, thanks to heating at depth. At that depth the boiling point of water, and therefore the temperature of steam, was 586 degrees Fahrenheit. The steam pressure gradient was 0.05 psi/foot. Applying more heat, by year 1979 the water/steam boundary had risen to 1750 feet below water table. But the pressure in the steam at the water/steam contact is determined by the pressure of the trapping water. At this lower pressure the boiling temperature is lower (493 degrees), and the steam is less dense.

How can steam lose temperature and pressure when it is being heated? That doesn't seem right! Answer chemistry and physics conspire. The temperature might remain high, superheating the steam in lower reaches of the reservoir, but at the water/steam contact the steam temperature would equal the boiling point of water at the depth of the water/steam contact. Steam behaves according to the reality shown in Figure 30.

Pursuing this line of reasoning further, if a geothermal hot spot would cause the ground temperature to rise to almost 212 degrees Fahrenheit, only steam at atmospheric pressure could exist in the subsurface. Cold rain and snow would complicate annually.

Maybe this condition of subterranean steam at atmospheric pressure exists at Yellowstone National Park?

So much for the water phase, now for the volcanic activity. Not only was the steam unroofed on the north flank of Mt. St. Helens, but consequently, pressure restraining the deeper volcanism dropped too. The volcanic glob of magma which had pushed out the north side of the volcano 450 feet during the preceding two months suddenly lost some of its confining pressure when the initial landslide eased the load. Then part of the northern flank of the mountain was blown away by the underground steam explosion, which further reduced pressure on the volcanic core. Naturally, the nascent volcanism there responded by releasing hot lava, ash and gases into the pandemonium. The trapped steam played out within a couple minutes of the start of the explosion. There was a change of leadership, and release of hot new volcanic materials became dominant. Volcanic eruptions of March 1980 in Mt. St. Helen's cone intensified dramatically during the big eruption on May 18. Fine ash rained down as white obnoxious dust.

Mt. St. Helens is one of about 450 volcanos around the Pacific, referred to as the "Ring of Fire". Previous to Mt. St. Helens, the last major volcanic activity in the USA was at Mt. Lassen in California in 1915.

E. MITIGATION?

Could a steam explosion in a threatening volcano be mitigated?

Mitigation of a steam explosion sounds attractive. Eliminating the massive density flow of reworked volcanic debris and steam at 500 degrees Fahrenheit would substantially reduce damage and loss of life from a volcanic eruption. At Mt. Saint Helens the steam explosion acted as a trigger for magma/lava activity. Without such a trigger, mitigation of a steam explosion might even postpone a lava eruption for decades or even centuries.

It would seem that by bleeding off the steam trapped under a volcano, the threat of a steam explosion could be nipped in the bud. A drainage system could be installed safely, using existing technology, from a remote location on a ridge several miles distant

Technically, so far so good. The objects of mitigation would be twofold: 1) to keep the water/steam contact deep enough so any landslide on the volcano could not unroof the steam, and 2) to drain pressure in the steam, reducing its pressure to a harmless level.

But there is a catch 22. Maybe mitigation is not such a good idea after all. These two objectives are incompatible. You can have one, but not the other.

Draining the steam pressure will cause the steam/water interface at the top of the trap to rise. Steam pressure will be reduced, but at the same time the elevation of the water/steam contact will rise. The resulting shallower steam /water interface will increase the danger of a shallow, smaller landslide unroofing the steam.

If the steam pressure can be substantially reduced before a magma blob develops in the flank of the volcano (i.e. before a landslide can occur), that would be advantageous. On the other hand, do you want to take the increased risk of a smaller landslide uncorking a steam explosion?

Alternatively, water could be injected into the volcano above the steam, to try to lower the water/steam elevation.

Careful engineering would be the byword here.

Prior to starting field operations for mitigation, getting approval **in time** for an Environmental Impact Report and government approvals might be challenging.

F. MID-ATLANTIC RIDGE

If the theory proposed here in **Moho Motion** is correct, a significant amount of heat is released annually under the Mid-Atlantic ridge from the released **heat of liquefaction**. Could there be a hazard similar to Mt St. Helens?

The pressure of cold ocean water two miles deep covering the mid-Atlantic ridge also permeates down below the ocean bottom in fractures and porosity. For fresh water, the **critical point**, where steam and water are one and the same, occurs at a depth of about 7500 feet below the water table. Hot sea water discharging from hydrothermal systems on the sea bottom has been sampled and analyzed. The measured critical point of that sea water is about 765 degrees Fahrenheit at a depth below sea level of 9730 feet (Google).

In mid-Atlantic, two miles of cold sea water plus water in the porous basalt create pressure in the water greatly exceeding the critical pressure there. So steam cannot be formed there, no matter how hot it gets. Because steam cannot be formed, a steam explosion, a-la Mt. St. Helens, is not possible below the mid-Atlantic ridge. Furthermore, a landslide at the bottom of the Atlantic would only reduce the vertical rock frame pressure, leaving the overwhelming controlling pressure of the sea water unchanged.

Under the mid-Atlantic ridge the superheated water/steam is less dense than the overlying cooler sea water. This creates a leaky trapping condition of hot, less dense sea water finding difficulty of buoyantly escaping up into its colder, denser counterpart. The interface between hot and cold waters is subject to the pathways of easiest egress for the hot water. In any case, "black smokers" and other hot water discharges with temperatures up to 870 degrees Fahrenheit have been observed flowing up from the Atlantic Ocean bottom (Google), so localized leakiness is demonstrated.

This measured water temperature is getting close to the Currie point of 1060 degrees Fahrenheit. Escaping water heated by lava (2100 degrees) and congealing basalt (1400 degrees) will erase the magnetic orientation of the basalt wherever the water heats the basalt above the Curie point. Upon cooling, the basalt will assume the ambient magnetic orientation. It is little wonder that the magnetic stripes on the Atlantic sea floor are imperfect.

Nineteen

Shape of Volcanos

A. CONES AND FLOOD BASALTS

Major volcanos occur in one of two shapes. Best known is the volcanic cone. Distinctive and easily recognized, in California there are three: Mount Shasta, Mt. Lassen and Mammoth. The other form of lava eruption is flood basalt, where lava simply spreads out on the earth's surface, onshore or offshore, forming no cone or mountain. A good example is the Colombian basalts in Oregon and Washington. Buoyant lava is channeled into either one or the other of these configurations. The "choice" depends on the easiest route for the buoyant lava. Is it easier (requires less work) to force its way up cracks, or to create a vertical feeder tube to escape?

Buoyant salt provides an interesting analogy, as pointed out by Andy Bengtson. Salt domes are more common than salt walls. Salt pillows and salt domes are essentially circular or oval in plan view, whereas salt walls are linear. Nature dictates that a thick layer of buoyant salt will do the least work to seek an escape from under the heavy burden of thousands of feet of sediment. First a salt pillow will lift up the overburden. When the buoyancy differential increases sufficiently, it would take less energy (work) to punch up through the overburden than for the pillow to expand laterally. So a column of salt, narrower than the pillow, rises slowly. These columns are often capped like a mushroom with a jumble of sediments pushed ahead of the salt. Some oilfields are found alongside salt domes, where originally horizontal sand layers are now oil reservoirs draped up on the flanks of the dome. The vertical salt column, a mile in diameter, has forced its way in and compacted the surrounding sediments. Salt walls form when less energy is required to buoyantly force salt up along a pre-existing vertical plane of weakness. If the least principal stress in the surrounding sedimentary rock is significantly less than the intermediate stress, the resulting weakness will favor a salt wall oriented in the direction of the intermediate stress.

Lava, being much less viscous than salt, will travel up inside a somewhat smaller throat. For example on California's coast at Morro Bay there is a lineup of several volcanic promontories, each about a thousand feet in diameter. These used to be feeder tubes for volcanos, which have since been eroded away.

B. STRESS CONCENTRATION IN WALLS OF FEEDER TUBES

These vertical volcanic feeder tubes concentrate stress in the surrounding rock through which they punch. In the cylindrical wall of a volcanic throat the stress in the rock will increase as much as double the regional horizontal stress away from the throat. The explanation for this hoop-stress involves simple calculus, which, thankfully, will not be attempted here. But, have faith, the effect of an empty vertical hole in a horizontal steel plate, compressed equally north- south and east- west, will cause the circular stress within the wall of the hole to be double the regional stress. A deadly rock spall spontaneously firing off the wall in a deep mine is another result of stress concentration in the wall of a void in a stressed elastic solid.

Drilling oil wells sometimes results in "running shale", where shards of shale will flake off from the wall of the well. The shale did not get ground up by the drill bit, but rather the hole became enlarged exposing the shale only to the drilling fluid, and not the drill bit. The exposed shale was compressed by the hoop stress. Perhaps the shale chemistry was incompatible with the drilling fluid. Shards of shale just popped loose.

Although the bit is round, the shape of the cross section of a well drilled in the earth is often enlarged and elliptical. The long axis of the ellipse is in the compass direction of the intermediate principal stress in the surrounding rock. The wall of the well bore became unstable because regional stress was concentrated in the wall of the well being drilled. This exceeded the strength of the shale, so the shale failed.

The extreme example of increased stress, caused by a drilled hole, is in a hollow drive shaft. When you put the pedal to the metal of an older car, torque is transmitted by a hollow tube from the motor to the rear wheels. In the tube, the amount of compressional stress is equal and opposite to the tensional stress. In engineering speak, this condition is called "pure shear". If you should drill a small hole in the driveshaft (a very bad idea), the stress in the wall of the small hole will be 16 times the compressional / tensional stress in the rest of the driveshaft. This will cause an explosive failure of the driveshaft when accelerating up the first steep hill, because steel is not strong enough to withstand 16 times the design stress.

The point is that the geometric shape of a hole or void in a stressed elastic material locally concentrates and magnifies the regional stress.

Back to a volcanic throat, the stress in the wall of a conduit for lava apparently strengthens the wall of the feeder tube. Many active volcanic cones resist leaking out the side. Japan's beautifully symmetrical Mt. Fuji is a prominent example. The pressure of hot lava thousands of feet above surrounding plains only occasionally forces its way out the side of a volcano, to form a subsidiary cone on the flank of the mountain.

C. CRATER LAKES ARE ALMOST WATER TIGHT

In addition to the burst resistance of a volcanic feeder tube, volcanic crater lakes show that the rest of the cone has an internal structure which holds water. When you visit Crater Lake in Oregon, the lake level is only a few hundred feet below the low points on the crater rim. That mountain, like all volcanic peaks, is a pile of volcanic debris sloping away from the core. How do you keep a whole lake from leaking out to the side of the crater, in such permeable rock?

Perhaps the answer is that as the volcano gets taller, the accumulating weight above the core settles down within the mountain like nesting paper Dixie cups, pointy ends down. Slip surfaces separating these conical settling units may inhibit water flow. It's a wild idea, but there is a place to test it.

The youngest Galapagos Island, Fernandina, is a young mile-high volcano on the west side of the Islands. It appears that the west side of Fernandina slid west into deep water, leaving a partially exposed cross section of part the west side of the volcano. Checking it out would be a plum summer field project for a grad student, provided Ecuadorian cooperation and permissions could be obtained.

D. SUBMARINE SEAMOUNTS

Under the mid-Atlantic ridge most of the erupting lava flowed up linear fractures and joints (cracks) trending north- south.

What about submarine volcanic cones along the mid-Atlantic ridge?? A "lava-breeder" out on the edge of the pull-apart zone might still exist where north-south fractures had been completely healed. Finding no egress into fractures/joints, lava would form a horizontal "bubble" under the congealed basalt. After buoyancy in the expanding bubble grew critical, the lava would find that less work was involved in punching up through solid basalt, compared to increasing the diameter of the bubble. Voila! A submarine volcanic cone is born!

Perhaps the cone will rise up above the waves, where erosion will flatten the top. One of Admiral Hess' guyots will result. As the crust with guyots loses one mile of elevation as it slides west /east off the humped up Moho under the mid-Atlantic spreading centers, the water depth to the top of a guyot should depend on its distance east or west from the spreading center.

❖   ❖   ❖

Twenty

As the World
Goes Round

A. VIRTUAL TOUR

You deserve a break after all this serious scientific stuff. You are hereby invited to go on a virtual tour to the South Pole and Ecuador! The real thing might take you ten weeks, depending on waiting time for weather. And the dollar cost could run to six figures. But this virtual tour will cost you no more than a few minutes of your time, and no money. What a deal!

For virtual starters, go to your neighborhood hardware store and buy a red brick and a can of yellow spray paint. At Sport Chalet get a small carrying bag. Spray one end of the brick yellow. Pop the brick into the bag, wait until November, and head for LAX and take the Christchurch flight. From New Zealand we are off to the McMurdo Station by the Ross Ice Shelf in Antarctica, having convinced the National Science Foundation that we are serious scientists on a mission to the South Pole. They let us in! Next we arrange for a motor trip over the compacted snow highway to the Amundson-Scott South Pole Station. The speed limit for the 1000 mile trek is 20 miles per hour. You would prefer a chopper flight? Well, we are traveling tourist class. Live with it.

Now we wait for a relatively clear day so that we can take up a position directly on the South Pole. There we put the brick on the snow and orient it with the yellow end facing the sun. Summer sunlight will graze both sides of the brick. Now we sit down and watch the brick. We are in luck! The sun shines all day. During 24 hours your brick will slowly rotate. After 12 hours the fully illuminated end will be brick red, and then finally yellow again. We can testify that it rotates, and does not change location. Your brick looks so quiet and harmless just lying there.

Now repack the brick and rack up airline-miles enroute to Ecuador's largest city, coastal Guayaquil near latitude zero. A rental car will transport you 50 miles west to Salinas, a sleepy little resort town on the Pacific coast. Relaxed tourists, generous uncles with their nieces, colorful natives and such a picturesque setting offer a most pleasant and peaceful tableau. Take your brick from its protective case. Lie down on the beach facing the Pacific. Place your brick on the sand with the yellow end facing you.

Hmmmmm. Let's see, here at the equator it is about 25,000 miles around our earth,

JUMPIN' JEHOSOPHAT!!

You just realized your six pound brick is hurtling right at your head at more than 1000 miles per hour!!

Few major league pitchers can manage to hurl a baseball 100 miles per hour. Very few tennis professionals can serve 150 miles per hour. Hot laps at the Indy 500 are clocked at 220 miles per hour. The muzzle velocity of an intimidating 1847 Colt-Walker six shooter is a little over 800 miles per hour, when using the larger powder charge.

Energy increases by the square of velocity. The most famous energy equation is Einstein's $E=Mc^2$. It has the same dimensions as all energy equations: mass times the square of velocity. It's just that in Einstein's case the amount of mass is the ridiculously small amount of matter that is converted to energy, and the velocity is the speed of light. For your brick, the equation for energy is the brick's mass times the square of the brick's speed.

Here in Salinas your brick is aimed right at you faster than a speeding bullet (shades of Superman!). The only thing saving you from instant death is that you are also moving along with your brick at more than 1000 miles per hour as the earth rotates. But the brick looks so benign, so peaceful in this seaside setting, just like it did at the South Pole. And you cannot remember how you and your brick got so much kinetic energy. You just sat there in your cramped seats in a succession of aircraft.

Now your brick packs as much wallop as would all the bullets simultaneously exiting the muzzles of more than 300 1847 Colt-Walker pistols, using the larger charges, all aimed dead east!

Your virtual tour is over! Now, what did you learn?

B. EARTHQUAKES AND THE EARTH'S ROTATION

The explanation for your high-speed, kinetic energy at the equator is simple. In 1835 a French scientist, Gaspard-Gustave Coriolis, wrote a paper, earning him permanent fame for "Coriolis acceleration". When you are at the South Pole, and you want to fly to Guayaquil by the shortest route along a line of longitude, you cannot point the aircraft straight north. The pilot uses Coriolis acceleration to compute his flight plan. The pilot must crab to the east, in the direction of the earth's rotation. This crabbing takes more energy than a direct flight. That additional energy means a passenger has more kinetic energy when he arrives. It's a little like swimming directly across a river, but aiming upstream so you will land at your target.

Conversely, flying directly south from Guayaquil to the South Pole takes less energy because you will be using some of your equatorial momentum to help fuel your flight south, as you crab to the west.

How is this related to plate tectonics? The subducting oceanic crust is known by the locations of earthquakes around the edges of the Pacific. As the Pacific gets smaller, the excess oceanic crust within the basin has to go somewhere. The Benioff curtains of earthquakes dip away from the Pacific. These curtains dip down into the earth for as much as hundreds of miles. It is easy to conclude that these earthquakes are the death throes of excess brittle oceanic crust disappearing under the edge of the shrinking Pacific. Basalt breaks when it stressed beyond its strength. Some of these Benioff earthquakes are whoppers. What causes these powerful releases of energy?

Slabs of brittle oceanic basaltic crust several miles thick are subject to the laws of physics. When a slab subducts 300 miles below the surface near the equator, the radius of its daily rotation is reduced from about

4000 to 3700 miles. However, as proposed in **Moho Motion**, the material below the Moho, through which the oceanic crust descends, does not move. Following the laws of physics, the basalt "tries" to maintain its angular momentum in its daily rotation around the world, so it pushes east as it descends. Remember your tetherball days as a kid? As the length of the tether gets shorter, the speed of the ball increases.

Trapped in the channel of its descent, the oceanic crust responds to the eastward push by trying to transfer the push against the eastern wall of the encasing channel. This leads to a tug of war, pitting the viscous material encasing the channel (fortified by lag time) vs the strength of the oceanic crust. The crust loses, and the crust breaks. At depths of 125 to 450 miles, the very high confining pressure on the slab of subducting basalt ensures that earthquakes can be powerful.

C. USGS EARTHQUAKE RECORDS

Beginning in 1973 the United States Geological Survey has maintained a 24/7 record of all earthquakes worldwide. You can search the record online for data on each earthquake (location, time, depth to hypocenter, magnitude).

Deep earthquakes occur only in subducting oceanic crust (Benioff zones). OK, let's set the bar at earthquakes deeper than 125 miles. In the 36 year period 1973-2009 there were 142 earthquakes, with magnitudes 6.5 and greater, recorded at depths deeper than 125 miles, worldwide. Of course, 37 years is just a speck of time in a geological context. But 142 earthquakes is not a small number, so let's follow this up. The deepest hypocenter was 423 miles in 2007. The greatest magnitude was an immense 8.2 (!) in 1994. 53 % of the 142 earthquakes were at depths greater than 313 miles. 92 events in the southern hemisphere averaged 18 degrees south latitude. In the northern hemisphere there were 50 earthquakes, averaging 33 degrees north latitude. The most northerly event was at 54 degrees north.

From this very limited data base we can make two generalizations about the Benioff zones: 1) at depths more than 125 miles below the surface there are more big quakes deeper than shallower, and 2) these quakes favor the low latitudes near the equator.

These observations are consistent with theory proposed here in **Moho Motion.** In our rotating earth, subducting oceanic crust moves independently from the stationary surrounding material below the Moho. If both crust and mantle were subducted together as a package, there would be no earthquakes. Earthquakes occur when crust and surrounding material have conflicting flight paths. Stress builds up until the brittle material (crust) pops.

Another way to contemplate low-latitude subduction is to think of applying the Coriolis Effect in three dimensions, rather than just on the earth's surface.

D. EASTERN DEFLECTION DEEP IN BENIOFF ZONES

Subduction zones generally dip down to the east under the Americas, and down to the west on the west side of the Pacific. In both cases conserving angular momentum will tend to force the stationary walls of the channels containing the subducting oceanic crust to the east. Therefore the shape of the Benioff zones on either side of the Pacific should be systematically different. Under the west coast of North and South America, once the subducting oceanic crust clears the continental crust and straightens out in the mushy material below the Moho, the easterly dip of its descent should tend to flatten out with greater depth. On the western side of the Pacific, where the Benioff zones dip west, the dip of the zone should tend to steepen with greater depth. Data from earthquakes near Japan may be better than those record-ed elsewhere in the west Pacific. So, despite the northerly latitudes there, perhaps this steepening will be detectable below Japan. In Brudzinski's projections of Japanese hypocenters reproduced in Figure 7 herein, steepening down dip to the west is apparent, as is predicted by **Moho Motion**.

Approaching the north and south poles, the radius of daily rotation shrinks. Thus, subduction occur-ring in more northerly and southerly latitudes should be accompanied by less powerful earthquakes, and less easterly deflection of the channels of subduction.

E. EARTH PLUMES AND EARTH ROTATION

A popular concept in the field of plate tectonics is the existence of buoyant plumes of heated mantle rising up from the earth's hot core.

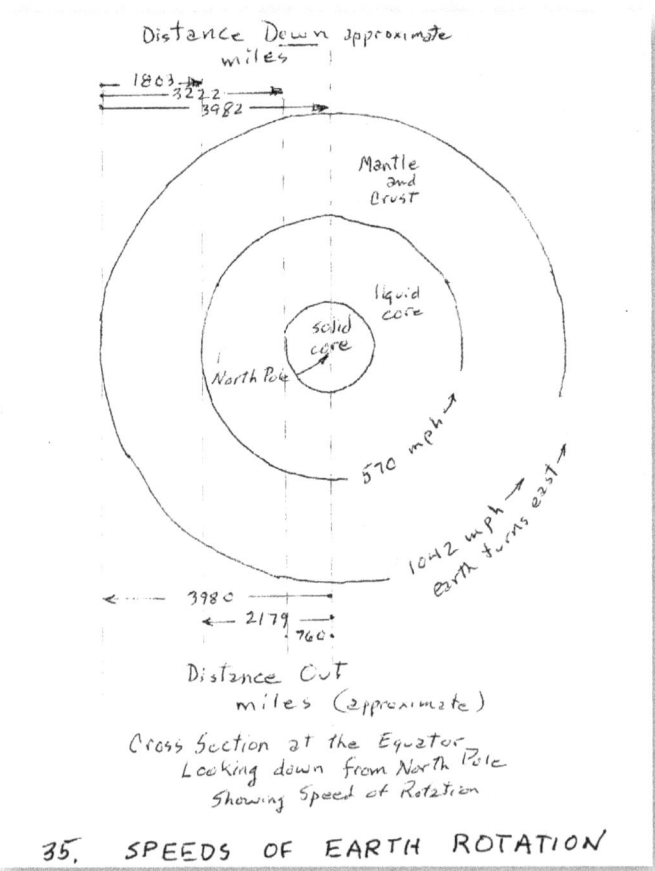

Figure 35. Looking down from the north axis of earth's rotation, this diagram shows the equatorial speed, in miles per hour, of the surface of the core, and of the earth's surface.

This diagram of the earth's interior is deduced from geophysical observations, and is reported in Wikipedia. Note that moving out from the core to the surface of the earth, the speed increases from 570 to 1042 miles per hour.

According to some current dogma, the mid-Atlantic ridge is underlain by an aligned string of individual plumes of buoyant mantle. As a plume get within about 200 miles of the surface, the rising column of mantle splits in half and bends from vertical to horizontal, half moving west, and the other half east.

It is said that the crust and upper mantle stick together as a solid plate (lithosphere). The theory is that a lower, mobile part of the mantle moves horizontally, dragging the lithosphere along. As the mantle spreads to the east and west it cools, and the density increases so it loses buoyancy and eventually sinks. This is the "treadmill" of mid-Atlantic spreading, as depicted in the next cartoon. (Have forbearance, please. Figure 3 is repeated here as figure 36, adding only a note of explanation.)

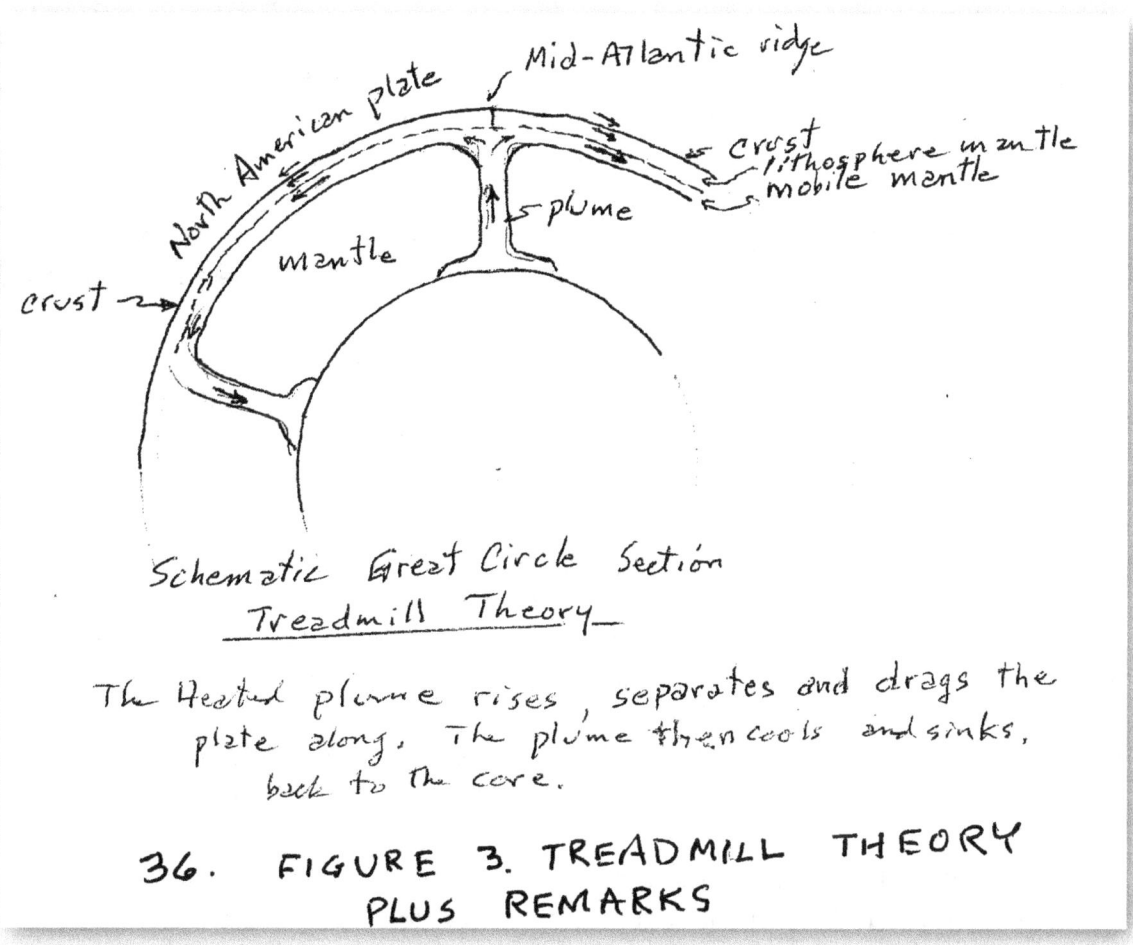

Figure 36. Cartoon of treadmill theory of plate movement. The lithosphere, including crust and upper 200 miles of mantle, moves at half the speed of the underlying 200 mile thick mobile mantle. The plume separates at 200 miles depth, half moving east, and half west.

Four problems with this treadmill scenario come to mind.

First, there is the effect of the earth's rotation. Consider the case **at the earth's equator**. According to the theory, a single plume will rise about 1600 miles until it splits into horizontal movement, east and west, about 200 miles below the mid-Atlantic ridge. The cooler mantle surrounding the plume is "stationary", in the sense that it is locked in with the earth's rotation.

On the earth's core the rotational horizontal speed is 570 miles per hour. The heated mantle buoyantly rises up from the core. As it rises the plume needs to go faster to keep up with the surrounding faster mantle. Starting at 570 miles per hour it must pick up the pace to 990 miles per hour at the split. The hot plume borrows energy by mixing with the surrounding faster cool mantle. Turbulence will result. The plume will be blown back to the west, like smoke in the wind.

The kinetic energy of moving from 570 to 990 mph horizontally requires that the rotating kinetic energy of the mantle in the plume increases 200 %, from the start at the edge of the core, up to the elevation of the split. Remember, energy varies as the square of the velocity ($E=Mc^2$),

So much for buoyancy at the equator. Now we head for one of the poles. Along the axis of the earth's rotation, **between the north and south poles,** a plume would rise vertically. No complications.

At **intermediate latitudes** it gets tricky. Buoyancy works radially away from the center of the earth, whereas rotation is around the axis of the earth's rotation. (Gets confusing, sort of like patting your tummy with one hand while applying circular massage to the top of your head with the other.) The net result would be westward lagging of ascending plumes, from maximum at the equator to zero at the poles.

So the **first** problem with the proposed scenario is that a) plumes will not be vertical, but will lag west greatly at the equator, but not at the poles, b) substantial turbulence and c) the plume of hotter mantle will be substantially diluted and cooled by having to borrow rotational energy to keep up with the surrounding cooler mantle.

Second, there is the timing. According to established dogma, below the mid-Atlantic ridge the mobile mantle drags the two plates east and west. Let's say the mobile mantle moves at twice the speed of the lithosphere that it is dragging along. Whereas the Atlantic got 2000 miles wider in 100 million years, the mobile lower mantle, moving twice as fast, moved 2000 miles east and 2000 miles west. Assuming that the speed of the rising mantle was the same as the speed of the horizontally spreading lower mantle, it would have taken about 80 million years for the plume to rise 1600 miles from the core to the split. This implies that an aligned string of pre-plumes formed on the surface of the core. These nascent plumes then coalesced as they rose simultaneously for 80 million years **before** getting near the earth's surface to form the mid-Atlantic ridge.

There are two illogical concepts here. The mid-Atlantic ridge started forming 80 million years before seafloor spreading started. Second, the required serendipity that all the plumes were lined up accurately on the edge of the core and they rose at the same rate and time, like puppets on strings.

Third, the apportionment of lava to the spreading centers. Rising mantle, being hot, is the hypothesized source for the magma in the mid-Atlantic spreading centers. But the mechanism for creating and containing frisky lava, for distributing the lava to each linear section of spreading, and for metering intermittent release of lava is incompletely explained in current theory.

Fourth, the buoyancy is excessive. We are used to thinking of buoyancy like a plastic ball in a swimming pool. Push it down two feet and it pops to the surface. But 1600 **miles** of buoyancy??

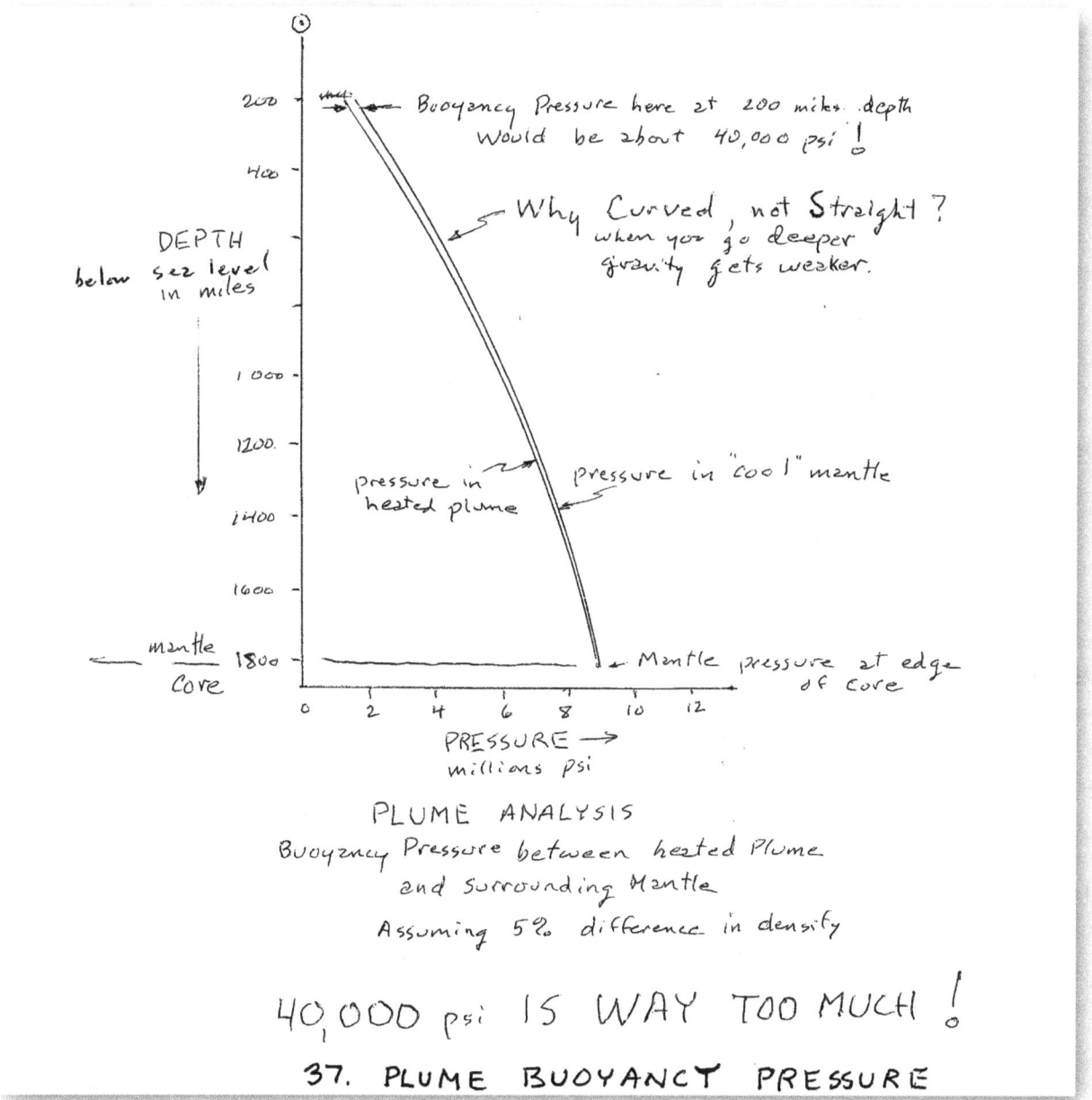

Figure 37-. This cartoon shows the pressure in the heated plume compared with the surrounding cool mantle, assuming a 5% difference in density. 1600 miles of buoyancy would amount to about 40,000 psi difference. Impossible. Turbulence would have long since destroyed the integrity of the plume.

At the edge of the earth's core, the pressure in the mantle, whether cool or being heated, would be the same. The heated mantle rises, buoyed in the surrounding stationary cooler mantle. Assume that density of the rising heated mantle is only 5 % less than in the surroundings cooler mantle. At the "splitting" depth of 200 miles, the pressure difference would be enormous, about 40,000 psi. Even if the difference in density were only 1 %, that would mean the buoyancy pressure at 200 miles depth would be 8000 psi.

Whether 8,000 or 40,000 psi at 200 miles depth, turbulence would have long since mixed the plume with the surrounding stationary cool mantle. A difference of only 40 psi (0.1% of 40,000 psi) would nudge mantle from hot towards cool. After all, the pressure in your auto tires is only 35 psi (twice atmospheric pressure), and that levitates your heavy Escalade. Turbulence, from juxtaposing hotter and cooler mantle at the same depth with pressure differences of hundreds of psi, would be lively, to say the least. Turbulence would dissipate the plume on the long trip up.

The turbulence would accentuate the western lag mentioned in the first problem (above).

F. EARTHQUAKES IN SUBDUCTING OCEANIC CRUST

In Chapter 10, **Late Breaking news,** Section B gives a plausible explanation for the two layers of hypocenters in Brudzinski's Figure 7.

Now let's add to that explanation the effects of the earth's rotation.

In Chapter 10 a case is made that the lower layer of hypocenters in Figure 7 is phantom. Also, the earthquakes in the subducting oceanic crust in the **western** Pacific under Japan apparently occur in the uppermost part of that slab.

So, how does earth's rotation fit into the explanation? In Figure 7 the shape of the layer of hypocenters under Japan is concave down. This means that the subducting slab does not remain straight, but instead it has been bent. In the **western** Pacific bending the rigid slab toward the east has caused it to fracture, and thus the Benioff hypocenters were generated.

The source of the earthquakes is the fracturing of rock in the **uppermost** part of the slab, as deduced from Figure 7. The subducting slab attempts to maintain its momentum, as its radius of daily rotation is reduced, so it presses east against the stationary plastic material below the Moho. The lag time in that viscous material cannot accommodate the pressure fast enough. On the bottom side of the bending slab the compressional head is evenly distributed, whereas in the **top** side of the bent slab there is relative relaxation, and the rock there fractures.

There is a different story for subduction under the **eastern** Pacific. Here the eastern push of the subducting slab presses the upper part of the slab against the material below the Moho, and the earthquakes are located in the relaxed **bottom** of the slab. The shape of the Benioff zone is concave upwards. This is opposite to the relationships in the **western** Pacific.

Under the high plateau in the South American Andes, the Benioff zone throws us a curved ball. The subduction path starts out normally for the eastern Pacific with decreasing dip to the east, but deeper it flattens out. Below that the Benioff path appears to roll over and steepen to the east. Puzzling!

We note that if the shape of the subducting slab is concave upward, the earthquakes occur near the bottom of the slab, and if the Benioff shape is concave downward the quakes are in the uppermost part of the slab. If this is true under the Altiplano, the hypocenters west of the inflection point in the shape of the Benioff zone will be near the bottom of the slab. Deeper to the east they will be located at the top. If all hypocenters are erroneously plotted as if they were located in the center of the slab, the apparent flattening will be exaggerated, perhaps greatly (depending of on the thickness of the subducting slab).

❖   ❖   ❖

Twenty-One

Floating on the Moho

It is proposed that the crust floats on the mushy material just below the Moho. Where can we go with that?

"Floating" means that on the Moho the pressure down is the same as, but opposite to, the pressure up. The vertical pressure down is lithostatic – all the air, ocean and crust resting on the Moho at that location.

The material below the Moho acts like a fluid in the long run. Short run, during passage of earthquake waves it acts like a solid. To analyze floatation, fluid characteristics, like viscosity, prevail.

The characteristic of floatation enables solutions to geologic problems via pressure-depth graphs. It's a little like a form of algebra. On a graph "floatation" is described as the intersection of two lines, the pressure up and the pressure down. Estimating crustal density and the depth to the Moho can be facilitated. Furthermore, errors in the model may be minimized by tweaking the assumptions and repeating the graphical solutions until the models are internally consistent. Of course, these graphical solutions, arrived at with a light table and a straight edge, are at the mercy of trembling hands. A computer could do the job and create graphs. However, the assumptions are less reliable than drafting accuracy, and a hands-on overview may prevent dumb mistakes from creeping in.

The head of the material below the Moho may be essentially constant over enormous distances. The undefined mechanism controlling the head below the Moho creates differences in head. The one-mile difference in head of the Moho from the mid-Atlantic ridge to the adjoining abyssal plain is enough to power movement of the North American plate. It may be that the head of the material below the Moho does not change much across the North American plate, from the Atlantic's abyssal plains to the Pacific. It may be that the density of the material below the Moho may also remain essentially constant. These super-simplifications may or may not hold water. Mother Nature is not one for uniformity. Even at the molecular level, there are exceptions.

In Figure 38 the objective is to use a pressure-depth graph to estimate the depth to the Moho under mile-high Denver, Colorado. Accuracy depends on the assumptions used in the graph. The first assumption is the most questionable, i.e. the head of the material below the Moho. This mega-assumption is that there are no changes in head of the material below the Moho under the North American plate from the Atlantic abyssal plain to far-off Denver. In Figure 38 the head of the material below the Moho remains the 3.1 miles subsea as graphically determined in Figure 23. The graphical solution is simple – just connect the gradient lines and where they intersect is the answer.

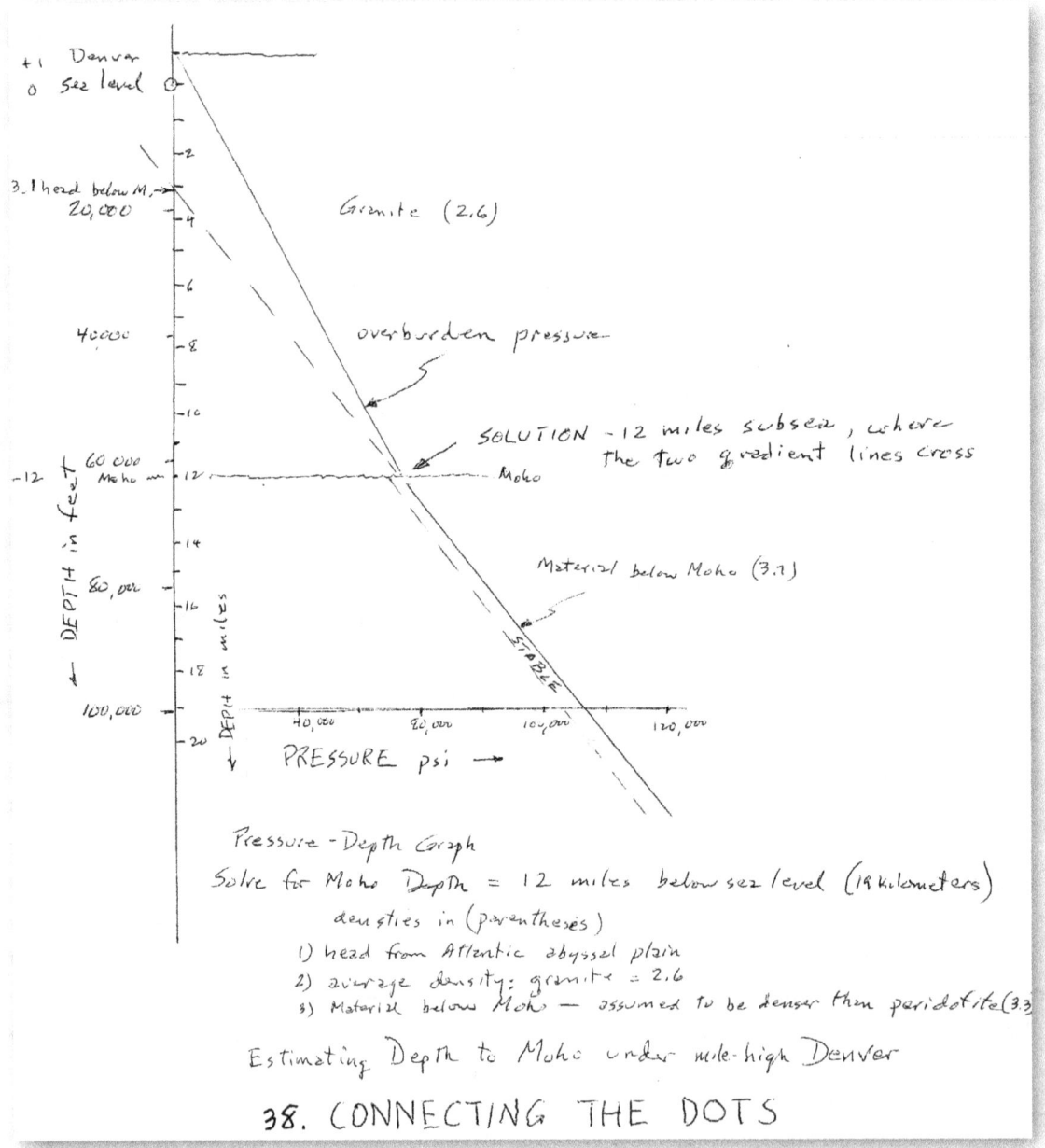

Figure 38. Connecting gradients to estimate the depth to the Moho below Denver, Colorado. Assumptions: 1) head of the material below the Moho, 3.1 miles subsea (same as Atlantic abyssal plain). 2) density above Moho: 2.6 (granite). 3) density of the material below the Moho 3.7. 4) elevation of Denver: 1 mile. Result: 12 miles subsea to Moho.

This solution is the product of a number of assumptions. To use this graphical system with intellectual honesty, the assumptions and solutions need to be massaged until they are internally consistent. Many computer runs of models, using different assumptions, might be needed to arrive at a best fit.

One of the assumptions is the density of the material below the Moho, which is estimated based on the densities of oceanic basalt (2.9) and peridotite (3.3). Oceanic basalt at density 2.9/3.3 "floats" on the material below the Moho, so a heavier density of 3.7 is arbitrarily assigned to the material below the Moho.

Whoops!! The United States Geological Survey's estimate for Moho depth near Denver from Figure 4 is 40 kilometers. Figure 38 estimates 19 kilometers. Big difference! Someone should go back to the drawing boards!

Who, me??!

Checking out the heat of liquefaction

Above and below the Moho the materials have the same chemical content. But they are in different states, with and without the **heat of liquefaction.** Could this be investigated in a high-pressure, high-temperature lab? 50,000 psi, 1300+ degrees Fahrenheit and lots of time would be a challenge and a half.

❖   ❖   ❖

Twenty-Two

Debunking:
"House of Cards?"

In Chapter 12 the **Moho Motion** theories were developed under the heading: "House Of Cards?" Now you have heard the pitch, and hopefully the arguments presented make sense. But "making sense" is a long way from proving validity. Real proof for disputable points may never be achieved -- much of the subject matter concerns deeply buried conditions which are only imaginable by man. But it is possible to reduce the areas of contention by gathering data, investigating and testing. Following are tests and avenues of study which could be carried out by investigators. Hopefully the results would bolster concepts proposed in **Moho Motion**, and the suspicion of a "House of Cards?" could be laid to rest. What follows is in no particular order.

A. EVALUATE BRUDZINSKI'S "LOCAL CATALOG" ILLUSTRATION FOR UNKNOWN REFLECTOR

In Chapter 10 the conclusion is expressed that acoustic chicanery is responsible for the second, lower band of hypocenters in Figure 7. Here is a simple test. It is assumed that the "picks" for the first arrivals of sound at the seismographs are automatically selected by a computer program. Devise a new program to eliminate all hypocenters occurring during the 10 seconds following any event plotting in the upper band of hypocenters.

Project the remaining hypocenters onto a cross section as in Figure 7. If this eliminates or substantially reduces the lower band of hypocenters, the culprit has been discovered. A reflector parallels the upper belt of hypocenters.

This reflector could be the Moho at the base of the subducting slab.

The eliminated time span between the upper and lower hypocenters in Figure 7 is the time it took for a roundtrip for sound from the hypocenter down to the reflector and back to the depth of the hypocenter. The delay times for rejected picks may be used to estimate the thickness of the subducting oceanic crust.

Earthquakes in the subducting slab in the western Pacific may be confined to the upper part of the slab. Brudzinski's upper band of hypocenters in Figure 7 seems to fit this hypothesis.

B. GET MORE DETAILS OF BENIOFF ZONES

Evaluation of the United States Geological Survey earthquake records suggest that the earth's rotation is related to the Benioff curtains of hypocenters. Suggestion: bring **Moho Motion's** tabulation of deep earthquakes up to date, and get into the details. Where were these deep earthquakes? Can the initial orientation of fault motion be added to the study? By lowering the thresholds (starting shallower, lower Richter numbers) can more detail be won?

C. SYSTEMATIC EASTERN DRIFT OF BENIOFF ZONES: WEST PACIFIC VS EAST PACIFIC

The shape of the Benioff curtains may offer proof whether subducting basalt slabs are systematically deflected to the east, as slabs slide deeper. Benioff zones in the western Pacific may steepen with depth, whereas those around the eastern Pacific may tend to flatten. If this is the case, then it is highly likely that the material below the Moho is stationary, whereas subducting oceanic crust moves independently. Piggy-back this study with the preceding item B.

D. MEASURE HEAT OF LIQUEFACTION IN LAB

The **heat of liquefaction** is basic to **Moho Motion's** theory. Try to measure it in the lab. Here is a plan. First core peridotite below the water table. Machine fragments of the core(s) to precise plugs 1 inch in diameter and a few inches long with square ends for a snug fit. Make numerous plugs so slow experiments can be carried out simultaneously. Prepare pressure chambers for cylindrical plugs, one end fixed and the other fitted with a piston. Install a very low wattage heater. Install electrodes at both ends for direct current pulses through the plug, to simulate natural electric flow in the earth (replace "other energy"). Install temperature and pressure sensors plus a detector for movement of the piston. Dust the peridotite plugs with lubricant and place in chambers. Pressure up the pistons to 50,000 psi in the peridotite and maintain constant pressure. Heat individual pressure chambers to 800, 1000, 1200 and 1400 degrees Fahrenheit. Vent any steam. Wait until temperature equalizes. Pass pulses of direct electric current through the plug. Commence slow heating and record temperature, pressure and movement of the piston.

The transition from peridotite to material below the Moho should be indicated by a flattening of the time/temperature curve, as **heat of liquefaction,** including "other energy", is restored. The temperature of the transition is the temperature matching the flattening of the time-temperature curve. At the same time the piston should move, indicating that the density is increasing. For those experiments which demonstrate that peridotite is being transformed to the material below the Moho, continue the heating and electrical shocking until transition is complete. Then lower the pressure and measure the amount of heat released (the **heat of liquefaction**). This may sound expensive, but it could probably be done for much less than the cost of coring below the Moho. If **heat of liquefaction** is confirmed by this experiment, momentum for coring below the Moho might lose steam, and serious money could be saved.

E. REFRACTION SEISMIC AROUND MID-ATLANTIC RIDGE

Is **Moho Motion's** structure of the mid-Atlantic spreading center correct? The thickness of oceanic crust under the Atlantic is poorly known. Refraction seismic surveys, with both geophones and source of sound on the sea bottom, probably have the best chance of achieving good results. Refraction might also prove whether the velocity of sound is faster in the north-south direction in the zone of spreading, and whether spreading centers under the mid-Atlantic are wider at the bottom, as proposed in **Moho Motion**.

F. ANALYZE MID-ATLANTIC EARTHQUAKES

Analyze the records of earthquakes in the mid-Atlantic spreading centers for the directions of initial slip. In **Moho Motion** it is proposed that these faults trend north-south, dip east and west, and are gravity faults.

G. CHEMISTS PLEASE FIND ANALOGIES

Chemists: please search for chemical processes analogous to **Moho Motion's** proposal that a **heat of liquefaction** separates chemically equivalent oceanic basalt from the denser material below the Moho. A good analogy would sure help sell validity here!

H. CONTOUR TOPS OF GUYOTS ON MID-ATLANTIC RIDGE

Contour the water depths to the crowns of Admiral Hess' guyots over the northern part of the mid-Atlantic ridge. Selecting the northern part should minimize the excessive lateral movement across the transform faults enabling the Bulge of Africa. Try to line up the zones of spreading by eliminating the offsets of all the transform faults. The resulting map pattern "should" be a 100 mile-wide, mile-high rise, tracking the ridge. This would prove gravity sliding off the mounded Moho under the mid-Atlantic ridge.

I. ANNUAL HEAT INPUT AND LOSS UNDER MID-ATLANTIC RIDGE

Measure sea floor water temperatures and compute a heat balance for a segment of the Mid-Atlantic ridge. Compare annual heat loss into the cold Atlantic waters to heat introduced into the spreading centers under that section of the ridge. Heat added annually is the release of the **heat of liquefaction** from new basalt separating the two plates. If the computed heat-in balances the measured heat- lost, then **Moho Motion** is on the right track

J. DRILLING PROBLEMS BELOW THE MOHO

Pursuing the Holy Grail of coring the rock below the Moho has a lot of pent up momentum among earth scientists. Come hell or high water, this project probably will survive. So when this drilling is undertaken, certain technical problems, which may be encountered, may substantiate **Moho Motion's** ideas.

Drilling requires that a special fluid be pumped down the hollow drill pipe to lubricate and cool the bit. This fluid also keeps the hole clean and returns the cuttings up to the surface. Drill cuttings from the material below the Moho will contain a great deal of heat energy which will be released into the fluid as it moves up the hole outside the drill pipe. The resulting increase in temperature of the drilling fluid may exceed the tolerance of the drilling equipment. **"Too hot!!!"**

In an effort to control a sudden increase in temperature, the fluid may be circulated without drilling deeper. Encouraged by a slow decrease of temperature, this circulation might go on for days. Meanwhile the viscous nature of the material below the Moho may cause the hole to deform, trapping the drill string below the Moho. **"Stuck pipe!!"**

If a representative core is cut below the Moho, and it is hoisted up inside the drill pipe, confining pressure on the core will be reduced, and the core will become incandescent lava. There goes your Holy Grail! **"Consternation!!"**

K. CHECK LOCATIONS OF VOLCANOS
   ## IN THE PACIFIC RING OF FIRE

In **Moho Motion,** Chapter 13, item S, it is concluded that volcanos in the Ring of Fire occur above the second bend of the subducting oceanic crust, and the triggering mechanism for eruptions is the same as for the mid-Atlantic ridge. Around the Pacific there are a number of deep reflection seismic lines perpendicular to subduction zones. Reflections may be recorded down to Moho depth. When nearby volcanos (The Ring of Fire) are projected onto these lines, one can see whether these volcanos are centered above the second bend of the subducting oceanic crust (figure 2).

Confirmation will strengthen **Moho Motion's** theory that volcanism of the mid-Atlantic ridge and in the Pacific Ring of Fire share the same triggering mechanism.

So, there you have it! Several new ideas and theories stand ready for the testing ingenuity of the world's geoscientists, especially you young ones who are not so beholden to established dogma.

About the Author

Jim's Norwegian parents, Dad from Horton and Mom from Hamar, met in Berkeley, California, where Jim was born in 1928. Jim grew up in Southern California, and continued his education there at Caltech, where he received a Bachelor of Science degree in 1950, majoring in geology. Skiing was a big attraction, so it should come as no surprise that the next stop was the University of Innsbruck, Austria. Earning a Ph. D. in 1953, his thesis was geological mapping around the Marmolada, the tallest peak in the Dolomites in northern Italy.

Graduation was quickly followed by marriage in Innsbruck and induction into the U.S. Army. The two year stint included basic training at Fort Ord, and photo mapping of Guadalcanal while serving in the Presidio of San Francisco.

As a raw recruit in the oil business, Jim's first job was in the Exploration Department of Standard Oil of California (now Chevron). He explored for oil in the three major onshore oil basins of California. He was Socal's "bird dog" on a seismic survey on the city streets of Los Angeles. Taking the State of California's tests, he became a registered petroleum engineer.

He switched employment in 1964 to Occidental Petroleum in Bakersfield. Occidental's first international venture was Libya, where Jim, wife, three daughters, dog and cat were Oxy's first overseas family in 1966. By the end of 1968, when he left Libya, Oxy's oil production there averaged 600,000 barrels per day, so Oxy had moved up to the big league. (And in September, 1969, along came Colonel Ghadafi!)

Back in Bakersfield, Jim worked exploration for Oxy in the eastern hemisphere. The next big venture was bidding for North Sea exploration blocks in the United Kingdom. Oxy acted as operator for a group of companies, and once again the small Oxy team was lucky and successful, and big oil production followed the discovery of the Piper and Claymore fields.

After twenty years with Oxy, Jim became a consultant in international petroleum exploration and production.

Jim became intrigued in finding a better explanation for the mechanics of how the Atlantic Ocean was getting wider. A quest lasting a quarter century resulted in this tome. Building a case for a new interpretation of the facts led from one new insight to another.

C. James Blom

203 Fairway Drive
Bakersfield, CA 93309

cjamesblom@msn.com

Author's Note

The ebook version of **Moho Motion** was originally published through Amazon in May 2015. Naturally, a few new ideas popped up while polishing the text for this paperback version, so the delay was worthwhile.

One new idea was that the **same** trigger mechanism for basaltic volcanism (releasing the **heat of liquefaction** when vertical pressure on the Moho is reduced) exists for the mid-Atlantic ridge, the Pacific Ring of Fire and for flood basalts whether on land or under oceans.

Thanks to the earth's rotation, subducting oceanic crust bends and fractures as the slab moves deeper in the stationary material below the Moho. Earthquakes in subducting slabs of oceanic crust in the western Pacific probably occur in the uppermost part of the slab. In the eastern Pacific, earthquakes probably favor the lowest parts of the slab.

Young earth scientists: if this unorthodox geological presentation of unorthodox geological ideas causes you to think again, and investigate these concepts, I have achieved my purpose.

And **laymen**: I hope you enjoyed and benefitted from playing the part of amateur earth scientists.

Halloween 2015

www.ingramcontent.com/pod-product-compliance
Lightning Source LLC
Chambersburg PA
CBHW080813180526
45168CB00006B/2425